Quantify!

Quantify!

A Crash Course in Smart Thinking

Göran Grimvall

The Johns Hopkins University Press
Baltimore

Printed in the United States of America on acid-free paper
9 8 7 6 5 4 3 2 1

The Johns Hopkins University Press
2715 North Charles Street
Baltimore, Maryland 21218-4363
www.press.jhu.edu

Library of Congress Cataloging-in-Publication Data

Grimvall, Göran.
Quantify! : a crash course in smart thinking / Göran Grimvall.
p. cm.
Includes bibliographical references and index.
ISBN-13: 978-0-8018-9716-0 (hardcover : alk. paper)
ISBN-10: 0-8018-9716-5 (hardcover : alk. paper)
ISBN-13: 978-0-8018-9717-7 (pbk. : alk. paper)
ISBN-10: 0-8018-9717-3 (pbk. : alk. paper)
1. Scientists—Psychology. 2. Creative thinking. 3. Lateral thinking.
4. Mathematical analysis. I. Title.
Q147.G75 2010
507.6—dc22 2010007501

A catalog record for this book is available from the British Library.

Special discounts are available for bulk purchases of this book. For more information, please contact Special Sales at 410-516-6936 or specialsales@press.jhu.edu.

The Johns Hopkins University Press uses environmentally friendly book materials, including recycled text paper that is composed of at least 30 percent post-consumer waste, whenever possible. All of our book papers are acid-free, and our jackets and covers are printed on paper with recycled content.

Contents

Preface

Being scientifically literate means being versed in the language of science and technology. In that world, quantification plays a central role. It could be about the amount of carbon dioxide in the atmosphere, the fuel consumption of cars, the memory capacity of hard disks, the limit values for noise in the workplace, or our own body mass index. Similarly, news about economy and sports is filled with data. An informed citizen must be able to interpret numbers and graphs. But here lie several difficulties.

A message may be completely misunderstood if one is not familiar with important units — for instance, knowing the difference between power expressed in kilowatts (kW) and energy expressed in kilowatt-hours (kWh). Limit values for radiation exposure, or the passenger capacity of an elevator, may give a false impression of sharp boundaries between what is dangerous and what is without risk. All measurements have uncertainties, and a difference from one measured value to another must be interpreted with great caution. Is it something to be taken very seriously or is it a statistical fluctuation?

These are just a few of the aspects covered in this book. The purpose is to illustrate how scientists and engineers think about things that can be expressed through numbers — sometimes very accurately and sometimes with large uncertainties. The examples are so general that the book has something to offer both those who have almost no education in science and technology, and those who are active in the field.

This book is the result of the author's lifelong efforts to unveil some of the general modes of thinking in science and technology, as a physics professor at the Royal Institute of Technology in Stockholm and as a frequent contributor to Swedish mass media. Of course, many colleagues have been helpful in this endeavor. In particular, I want to thank Lars G. Larsson, Torbjörn Thedéen, and Anders J. Thor for comments on the text.

Quantify!

1 Numbers

1.1 Numerical Literacy

What is meant by one billion liters, or one exajoule, and how Europeans write the date 9/11.

Babylon, Babble, and Billion

"Let us go down and confuse their language so they will not understand each other." According to the Bible, this is what happened to the ancient city of Babylon.[1] In sharp contrast, today's scientists use the same language worldwide. For instance, while animals and plants have names that vary with the language, their Latin names are unique. It may seem surprising then that the names of numbers cause so much confusion.

How many cubic meters make one *billion* liters? The answer may depend on where you live. In American English, one billion is the same as one thousand million. In most European languages, however, one billion, or its equivalent word, is the same as one million million. This difference has caused numerous errors and misunderstandings. Similarly, the word *trillion* in American English means one million million, but in Europe it usually means one million million million. Table 1.1 illustrates this difference for some languages. There is a word in European languages, *milliard* or its equivalent, which means one thousand million. In 1974 the U.K. government adopted the American usage in official documents, but some people still adhere to the old British meaning. If one wants to be absolutely clear, one can say "thousand million" instead of "billion." Similarly, in Spanish one can say *mil millónes*. The International Organization for Standardization (ISO)[2] does not accept the common use of parts

Table 1.1. Names of powers of 10 in various languages

Power	*US English*	*UK English*	*French*	*German*	*Spanish*
10^3	thousand	thousand	mille	Tausend	mil
10^6	million	million	million	Million	millón
10^9	billion	milliard	milliard	Milliarde	millardo
10^{12}	trillion	billion	billion	Billion	billón
10^{15}	quadrillion				
10^{18}	quintillion	trillion	trillion	Trillion	trillón
10^{21}	sexillion				
10^{24}	septillion	quadrillion	quadrillion	Quadrillion	cuatrillón

per billion (ppb) for two reasons: the meaning of *billion* is ambiguous, and *parts* is also ambiguous because it could refer to volume, mass, number, or some other measure.

In American English, explicit names are given to powers of 10 in the sequence

$$10^{3+3n}$$

for $n = 1, 2, 3, 4, 5$, which is read as million, billion, trillion, quadrillion, quintillion, followed by nonunique names. In most European languages, names are instead given to the sequence

$$10^{6n}$$

for $n = 1, 2, 3, 4$, which in this case is also read as million, billion, trillion, quadrillion, and so on.

Many Asian languages have names for the sequence 10^{4n} instead of the sequences of 10^{3n} and 10^{6n}. For instance, there are Chinese characters for 10, 100, 1000, 10^4, 10^8, 10^{12}, and so on. Figure 1.1 shows the characters for 10, 10^4, and 10^5. We see that the symbol for 10^5 is that for 10 followed by 10^4. The Japanese and the Chinese systems are very similar. The Indian numbering system follows the sequence 10^{1+2n}. The difficulties caused by the differences between the European and North American systems with names based on 10^3

Fig. 1.1. Chinese characters for 10, 1000, and 10 000

and 10^6, and Asian systems with names based on 10^2 and 10^4, should not be underestimated. Although it is no problem when one has time to sit down and write out the powers of 10 explicitly, many people can tell of misunderstandings in ordinary conversations.

Prefixes

Words like *megabyte*, *terawatt*, and *nanoscience* are common, even in ordinary newscasts. The engineer may talk about *picofarad*, *exajoule*, and *femtosecond*. All such names contain prefixes, which denote powers of 10 from 10^{-24} to 10^{24} (table 1.2). Some of the prefixes are derived from words for numbers. For instance, *femto* and *atto* come from the Danish *femten* (15) and *atten* (18). *Pico* comes from the Spanish word *pico* ("small"). *Nano* is not a word for "nine," as many people think, but comes from the word for "dwarf" (Latin *nanus*, Greek *nanos*). Several prefixes originate from Greek words related to size: *mikros* ("small"), *megas* ("huge," "powerful"), *gigas* ("giant"), *teras* ("monster"). Sometimes the word *terawatt* is incorrectly written and pronounced as *terrawatt*, but the prefix has nothing to do with the Latin word *terra* ("earth").[3]

Prefixes are written with capital letters when they refer to numbers larger than 10^3. Those that denote smaller numbers are written with lowercase letters. The symbol k for *kilo* ("one thousand") is often incorrectly written as K. For instance, it is not unusual to see signs with texts like "Hotel 2 Km" or "Maximum hand baggage 8 Kg." If we follow the international standards of units and their symbols, these two signs would be read aloud as "Hotel 2 kelvin-meter" and "Maximum hand baggage 8 kelvin-gram" (the symbol for the temperature unit kelvin is K). In these two cases, it is of course impossible that the signs would be misinterpreted, but

Table 1.2. SI prefixes and symbols for powers of 10

Factor	*Name*	*Symbol*	*Factor*	*Name*	*Symbol*
10^{24}	yotta	Y	10^{-1}	deci	d
10^{21}	zetta	Z	10^{-2}	centi	c
10^{18}	exa	E	10^{-3}	milli	m
10^{15}	peta	P	10^{-6}	micro	μ
10^{12}	tera	T	10^{-9}	nano	n
10^{9}	giga	G	10^{-12}	pico	p
10^{6}	mega	M	10^{-15}	femto	f
10^{3}	kilo	k	10^{-18}	atto	a
10^{2}	hecto	h	10^{-21}	zepto	z
10	deka	da	10^{-24}	yocto	y

Table 1.3. Names and symbols for prefixes of binary powers

Factor	*Name*	*Symbol*	*Factor*	*Name*	*Symbol*
$2^{10} = 1024$	kibi	Ki	$2^{40} = 1024^{4}$	tebi	Ti
$2^{20} = 1024^{2}$	mebi	Mi	$2^{50} = 1024^{5}$	pebi	Pi
$2^{30} = 1024^{3}$	gibi	Gi	$2^{60} = 1024^{6}$	exbi	Ei

we will see examples in this book where mistakes with units have caused large economic losses and the risk of lives.

Food tables often speak of "calories." The food "calorie" actually is a kilocalorie — 1000 calories. The symbols are cal for calorie and kcal for kilocalorie. The calorie is an energy unit that is now largely obsolete. Most scientific works use joule (J) instead, with 1 cal = 4.1868 J, and many food packages give the energy in both kilocalories and kilojoules. The European Union formulated a directive which would ban such a use of two systems of units after January 1, 2010, but after strong protests the implementation of "SI units *only*" has been suspended.[4]

Most people interpret 1 kilobyte as 1000 bytes, 1 megabyte as 1 000 000 bytes, and so on. This is an approximation, and in information technology one generally uses the binary representation. Because 2^{10} byte = 1024 byte, and not 1000 byte, special names are given to the approximate powers of 10 (table 1.3). The ending *-bi* in the names is derived from the word *binary*, and pronounced "bee."

What Is the Point?

Consider the expression

$$\frac{1}{8}\text{ inch} = 3{,}175\text{ mm.}$$

It can be confusing—in Europe as well as in the United States and in any other country. According to the ISO, the decimal sign is written either as a comma or a point (a period), but in documents written in English the decimal sign is usually, although not always, written as a point. For instance, an exception is the English version of the ISO Standards, where the comma is used as the decimal sign. This book uses the point. In that way we avoid the confusion that may otherwise arise because of the unofficial convention in the United States to use the comma to separate groups of three digits. Americans would interpret the expression 3,175 mm as 3175 mm, but this is in conflict with the ISO rule. The 22nd Conférence Générale des Poids et Mesures in 2003 repeated that "the symbol for the decimal marker shall be either the point on the line or the comma on the line." Furthermore, it reaffirmed that when numbers are divided in groups of three in order to facilitate reading, "neither dots nor commas are ever inserted in the spaces between groups."

For monetary amounts, however, there is a risk of forgery; thus, a written character is required in every character position. Points, and not blank spaces, are then used to separate groups of digits. For instance, we would write

EUR 125.560.000

When the demand for security against forgery is greater than that for readability, the amount is written without digit separation:

EUR 125560000

If there is a need for additional protection against forgery, the monetary amount is supplemented by letters:

onehundredandtwentyfivemillionfivehundredand
sixtythousandeuros

A half-high dot is often (and correctly) used as a multiplication sign. In countries where the decimal sign is written as a point, a recommended alternative to the multiplication dot is a cross, as in 1/8 inch $= 3.175 \times 10^{-3}$ m.

The many different ways for writing dates may lead to serious mistakes, or at least to confusion. According to ISO 8601, the date December 11, 2001, can be written in any of three ways:

20011211, or 2001-12-11, or 01-12-11

In the United States this date would usually be written

12/11/01

Those who are used to the ISO standard may be uncertain if this is meant to denote December 11, 2001, or November 12, 2001, or perhaps November 1, 2012. It becomes even worse if one suspects that the writer in fact has tried to apply ISO 8601 rather than the American style, but has made the mistake of using the slash symbol (/) instead of a hyphen (-) to separate the groups of numbers. In that case the interpretation of 12/11/01 would be November 1, 2012. In some countries one often uses points, as in 11.12.01 or 11.12.2001, to denote December 11, 2001. To add even further to the differences in conventions, the European Union has decided that dates on food packages must conform to the style DDMMYY (day, month, year), without any symbols between the groups of digits denoting day, month, and year. However, the International Postal Union has adopted the ISO format YYMMDD. Finally, some people write December 11 and others write 11 December, but at least that would not lead to ambiguities (fig. 1.2).

Fig. 1.2. An unambiguous way to express a date

1.2 The Power of Logarithms

Why 3 is in the middle between 1 and 10, how Renard reduced the number of rope dimensions for hot air balloons from 525 to 17, and how taxation authorities can detect fraud.

Order of Magnitude

Science deals with objects and phenomena that can vary enormously in magnitude. The distance to the nearest star, Proxima Centauri, is about 4 light years or 4×10^{16} m, and the diameter of an atomic nucleus is about 10^{-14} m. The energy that the Earth receives from the Sun every second is more than one million times larger than what is needed for one year's use of a car. It is common to describe phenomena that span such wide ranges with a diagram drawn in a logarithmic scale. Figure 1.3 shows the mass of some organisms, from a bacterium to the largest animal on earth—the blue whale. The mass of a blue whale is larger than the mass of a bacterium by 12 orders of magnitude, meaning that the mass ratio is about 10^{12}.

When scientists say that two things differ by an order of magnitude, they often have in mind a difference by about a factor of 10, but the concept can also be used in a less precise way. Nanoscience deals with phenomena where it is natural to measure lengths in the unit nanometer, in which 1 nm = 10^{-9} m. This is in contrast to phenomena where lengths are naturally given in, for instance, the unit micrometer, with 1 μm = 10^{-6} m. If something has a length of 8 μm, some people may say that the size is of the order of a microme-

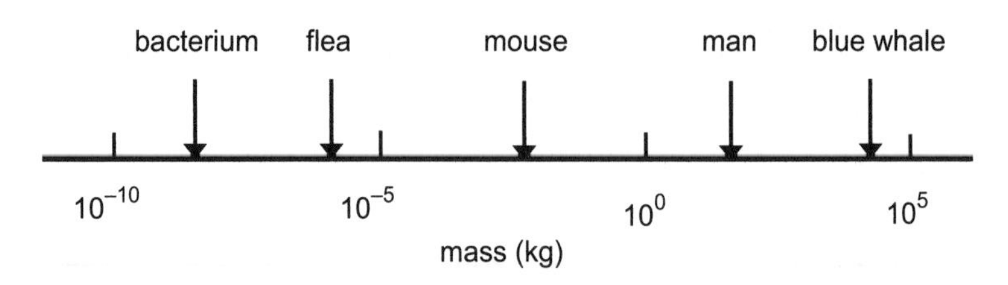

Fig. 1.3. The mass of some organisms, on a logarithmic scale

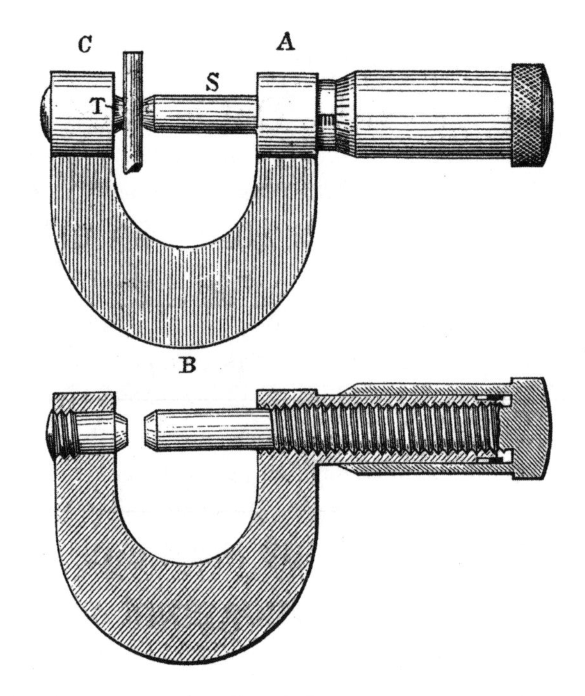

Fig. 1.4. A micrometer is a length and also an instrument used to measure lengths, typically in the interval 1–1000 μm

ter, while others may say it is of the order 10 μm. It depends on what one compares with and how precise one wants to be. *Micrometer* is also the name of a measuring instrument (fig. 1.4).

Astronomers express the apparent brightness of stars and other celestial objects in terms of their magnitude. As a reference level, the

Table 1.4. The apparent brightness of some celestial objects, expressed as magnitude

Sun	*Full Moon*	*Venus*	*Sirius*	*Spica*	*North Star*
−27	−13	−4.4	−1.5	1.0	2.1

North Star is given the magnitude 2.12. One step in the magnitude of a star means a change in apparent brightness by the factor

$$10^{2/5} \approx 2.512.$$

Table 1.4 gives the magnitude of some celestial objects. Because the sensitivity of the eye depends on the wavelength, a more precise definition is needed when the magnitude of brightness is used, for instance, in a study of photographs. Magnitude values *increase* as the star gets fainter, which may be contrary to what most people would think. For instance, one can barely see a star of magnitude 6 with the naked eye on a dark night. (Magnitude on the Richter scale is discussed in Chapter 2.)

Hot Air Balloons and Renard Numbers

The French army in the 1870s used no fewer than 425 different rope diameters for its hot air balloons. Then Captain Charles Renard devised a system that reduced the number to 17. He got the bright idea that the diameter should increase in steps governed by a certain *factor* rather than, for instance, by a certain width expressed in millimeters. His concept now has widespread use in technology. It is also used, approximately, in the denominations of coins and bills, as in the sequence of coins of 1, 2, 5, 10, 20, and 50 cents, 1 and 2 euros, in 16 of the 27 countries in the European Union (fig. 1.5).

The ISO standard defines four base series for Renard numbers, denoted R5, R10, R20, and R40. In the R5 series, each number is larger than the preceding one by the factor[5]

$$\sqrt[5]{10} \approx 1.584.$$

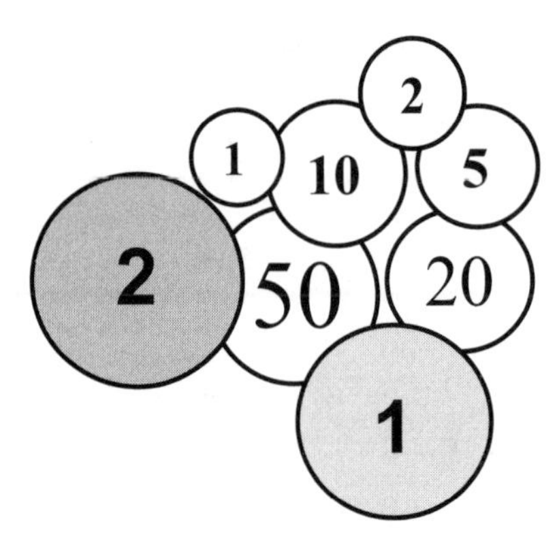

Fig. 1.5. A typical sequence of denominations for coins

For practical reasons, the numbers are slightly rounded, as shown in table 1.5 for R5 and R10. They take the same structure within every decade and are evenly spread when plotted on a logarithmic scale.[6] For instance, R5 also gives the number series 1000, 1600, 2500, 4000, 6300, and 10 000.

If you look at a catalogue of electronic equipment, you may be surprised to find resistors with resistance values that appear somewhat odd. For instance, a common value is 56 ohms. Why not 50 ohms, particularly since there are also 10-ohm and 100-ohm resistors in the catalogue? The explanation is that the number 56 belongs to the so-called E12 series, which has a pattern similar to that of the Renard series.[7] Each decade is divided into 12 steps, with an increase in each step by a factor of

$$\sqrt[12]{10} \approx 1.212.$$

The reason for this choice is that the resistances were once given with an uncertainty of 10 %. A nominal value of 100 thus could be as high as 110 in reality, or as low as 90. By choosing resistance values that increase (approximately) according to the E12 scale, one could cover any desired resistance within the 10 % uncertainty. For instance, one has

$$(\sqrt[12]{10})^{9/12} \approx 5.6,$$

which explains the resistance value 56 ohms.

Table 1.5. The numbers in the Renard series R5 and R10

R5	1.00		1.60		2.50		4.00		6.30		10.00
R10	1.00	1.25	1.60	2.00	2.50	3.15	4.00	5.00	6.30	8.00	10.00

Finally, note that scientists and engineers often say that the number 3, rather than 5 or 5.5, lies in the middle between 1 and 10. That is a practical approximation to the value

$$\sqrt{10} \approx 3.162.$$

In the same spirit we can say that the number in the middle between 1 and 2 is approximately 1.4, and not 1.5, because $\sqrt{2}$ differs from 1 by the same factor as 2 differs from $\sqrt{2}$. The ISO standard 216 for paper sizes uses the $\sqrt{2}$ aspect ratio, leading to the A4 letter size that is widely used in the world, although not in the United States.[8]

Finding Fraud in Figures

One might think that the digits 1, 2, . . . , 9 occur with about equal probability in a set of data giving, for instance, the area of lakes, the exchange rates of currencies, or the expenses paid by a company. This is usually wrong if one considers the *first* digit in each entry.[9] The number 1 is more frequent than 2, which in turn is more frequent than 3, and so on, up to 9. This frequency distribution has been used by the US Internal Revenue Service (IRS) to search for fraud in cases where numbers were thought to be freely invented.

The uneven distribution of the value of the first digit has been known at least since the second half of the nineteenth century, but credit is often given to the American physicist Frank Benford, who in 1938 published an extensive analysis. He noted that in books of logarithmic tables, the pages containing the logarithms of numbers beginning with 1 were more worn than pages for numbers beginning with 9. Before the existence of calculators, such tables of logarithms were used to facilitate the multiplication of numbers, in particular in

Table 1.6. Gross domestic use of energy, 2006, in the 27 EU countries

Country and EU abbreviation	*10⁶* toe	PJ	TWh
Belgium, BE	60	2500	700
Bulgaria, BG	21	860	240
Czech Republic, CZ	46	1900	540
Denmark, DK	21	880	240
Germany, DE	349	15 000	4100
Estonia, EE	5	230	63
Ireland, IE	16	650	180
Greece, EL	32	1300	370
Spain, ES	144	6000	1700
France, FR	274	11 000	3200
Italy, IT	186	7800	2200
Cyprus, CY	3	110	30
Latvia, LV	5	190	53
Lithuania, LT	8	350	98
Luxembourg, LU	5	200	55
Hungary, HU	28	1200	320
Malta, MT	1	38	10
Netherlands, NL	81	3400	940
Austria, AT	34	1400	400
Poland, PL	98	4100	1100
Portugal, PT	25	1100	290
Romania, RO	41	1700	480
Slovenia, SI	7	310	85
Slovakia, SK	19	790	220
Finland, FI	38	1600	440
Sweden, SE	50	2100	590
United Kingdom, UK	229	9600	2700

Note: Countries are ordered according to their names in their own language, with some names transliterated here; thus, the following countries fall out of the usual alphabetical order in English: Deutschland (Germany), Éire (Ireland), Elláda (Greece), Kýpros (Cyprus), Magyarország (Hungary), Österreich (Austria), and Suomi (Finland). The two-letter country codes (e.g., BE for Belgium) used by EU are according to the ISO Standard 3166-1, with the exception of EL (not GR) and UK (not GB) astronomy. The distribution of the first digit is sometimes called Benford's law or the significant-digit law.[10]

The following argument can give an intuitive feeling for Benford's law.[11] Consider a set of numbers x, which are randomly chosen in the interval between 1 and 10. All x between 1 and 1.999 have 1 as the first digit, all x between 2 and 2.999 have 2 as the first digit, and so on. The ratio of the largest and the smallest x is (almost) 2 in the first interval. In the next interval it is 3/2, followed by 4/3, and so on, up to 10/9. The widths of the intervals decrease as in a logarithmic scale. There is more "room" for a number between 1 and 2 than there is between 8 and 9.

Benford's law can be illustrated with data for the energy used in the 27 EU countries. Energy use is expressed in tonnes (tons) of oil equivalent (toe), an energy unit adopted by the Organization for Economic Cooperation and Development (OECD). The definition is 1 toe = 10^7 kcal, which closely corresponds to the net heat content of 1 tonne of crude oil: 1 toe = 41.868 GJ = 11.63 MWh. It is arbitrary which of these units we choose for energy use; the rule about the first digit should hold irrespective of that choice. The values in table 1.6 are rounded to one or two significant digits. Figure 1.6 gives the number of times a certain first digit occurs in any of the three columns in table 1.6, together with the expected number according to Benford's law.

The rule about the first digit fails if the data under consideration vary by less than about one or two orders of magnitude. For instance, in most countries many more people die at an age (in years) beginning with the digits 6 to 9 than at an age beginning with the digits 2 to 5. It makes no essential difference if we instead express the age in, for example, hours. A lifetime of 50 years corresponds to about 438 000 hours, and a lifetime of 90 years corresponds to about 788 000 hours. Thus, when the life span is measured in hours, it is more likely that the first digit is 5, 6, or 7 than 1, 2, 3, or 4.

The reason for the failure of the first-digit rule is that the life span of humans normally varies by less than a factor of 1.5, although the difference can be much larger in some cases. The factor 1.5 is, of

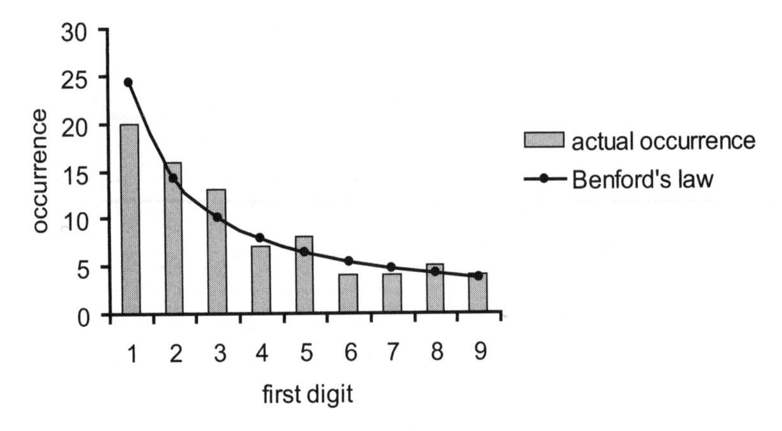

Fig. 1.6. The actual distribution of the first digit for the entries in table 1.6 and the distribution according to Benford's law

course, independent of the time unit used to express the length of life. Therefore Benford's law is not obeyed in any unit system used to describe the length of a life. Similarly, the energy values in table 1.6 reflect the size of a country. If we had considered the energy per capita, the numbers would lie in a rather narrow interval, and the distribution law for the first digit would not hold.

1.3 What Is Typical?

The typical height of a person, the typical number of people you meet during a long life, and the acceptable risk of being hit by a returning rocket.

The Height of an Adult

The height of an adult human being is typically 1 m. Many people would object to such a statement because they know that adults range in height from about 1.5 m to 1.8 m (5–6 ft). But the word *typical*, or *characteristic*, can have different meanings. If we put the human race on a logarithmic scale such as that shown in figure 1.3, which gives the size of organisms from bacteria to blue whales, it

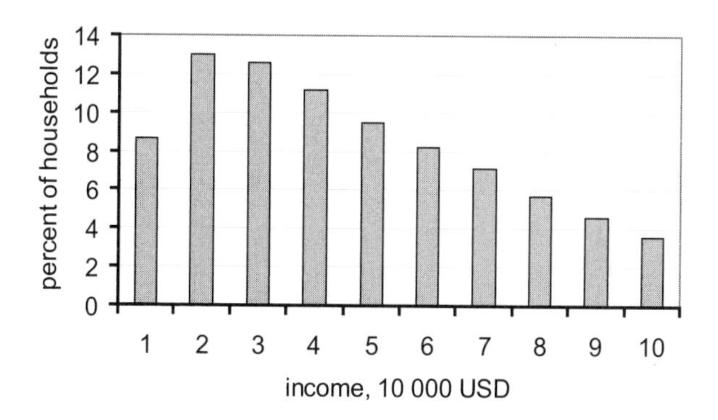

Fig. 1.7. The annual income per household in the United States, 2005, for intervals of $10 000, up to $100 000

would not make much of a difference whether we say that the height is 1 m or 1.6 m. But a designer of chairs would not accept such imprecision about people's sizes!

Figure 1.7 shows the distribution of household incomes in the United States in 2005 divided into 10 equally wide intervals from $0 to $100 000. Above $100 000 there is a very long tail, containing 12 % of all household incomes. What was a "typical" household income in 2005? The largest number of households fall in the interval $10 000–$20 000. We could also take the median income—that is, the value for which an equal number of households lie above and below, respectively. That gives a median income range of $40 000 to $50 000. Finally, we could take the average income, defined as the total income divided by the total number of households. This result depends significantly on the few very high incomes and in the United States is about 40 % higher than the median household income. In a society with a small income spread, any of these three alternatives could serve to define a "typical" or "characteristic" value.

If incomes are compared between countries, one must factor in differences in the structure of households. For instance, do many of the one-person households refer to students with low incomes? Do the elderly commonly live alone rather than with their children? Of

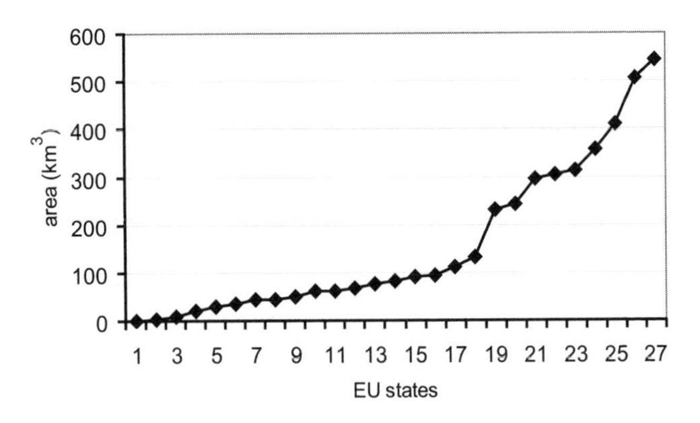

Fig. 1.8. The areas of the 27 EU countries arranged in a cumulative graph

course, the extent of social benefits like free or low-cost education and medical treatment must also be accounted for.

Here is another example, as a background to the concept of what is "typical." In 2007, the European Union grew to include 27 independent states. They vary in area from the two smallest (Malta, 300 km^2; Luxembourg, 2600 km^2) to the two largest (France, 544 000 km^2; Spain, 506 000 km^2). What is the typical area of a state in the European Union? It turns out that the 27 entries, spanning a range larger than three orders of magnitude, cannot be represented by a more or less bell-shaped curve such as that in figure 1.7. Instead, we plot the data as a *cumulative* graph (fig. 1.8).[12] If this curve had a wide plateau, we could identify that value as the typical size of EU states, but this is not what the figure looks like. Austria represents the median, in between 13 smaller and 13 larger states. Its area, 82 500 km^2, could perhaps be taken as a characteristic EU size. But if we define the phrase "within half an order of magnitude" as meaning "within a factor of $\sqrt{10} \approx 3.16$ or $1/\sqrt{10} \approx 0.316$," then only 16 of the 27 EU states have an area equal to that of Austria within half an order of magnitude. Therefore, we might even say that there is no typical area for an EU state.

Compare this with, for instance, the heights of adult humans. They certainly lie within half an order of magnitude from a mean of about 1.7 m, being within the interval 5.4–0.54 m (18–1.8 ft). It is obvious from our three examples—the height of a person, household income in the United States, and the area of EU countries—that the concept of what is typical or characteristic depends on the case one refers to. Nevertheless, even vague measures can have a useful meaning in a comparison with something that is very different. For instance, the area of EU countries is typically much larger than the area of the city of New York, although Malta is actually smaller.

Social Competence and Personal Encounters

The Earth has almost 7000 million inhabitants, speaking almost 7000 different languages. How many words in total are uttered by them, per second and right now? If that question is fired at you, requiring a quick answer, the number you give may be completely unrealistic. Of course, no one knows the precise answer, but given some time to think, we should be able to come up with a result that is not wrong by more than a factor of 10. Let us start with input data that we may later want to change. If one utters two words per second, the first three sentences in this paragraph would take a little more than 20 seconds. (You may check that this is reasonable.) If all people on our planet talk without interruption, around the clock, it would mean $2 \times 7 \times 10^9$ words per second.

Next, for how large a fraction of a day does one actually talk? That varies a lot, with cultural background, age, profession, and so on. Suppose that you and your friend discuss something intensely for an hour, or 30 minutes of talking for each of you. In a group of four, there is typically 15 minutes' talk per person. It should be very rare that an individual's effective time of talking (i.e., excluding short or long pauses) is less than three minutes or more than three hours per day. For the sake of simple calculations, let us take it to be 24 minutes, or 1/60 of a full day. Then a total of $(2 \times 7 \times 10^9)/60$, or about 200 million, words are uttered somewhere, every second. This is only a crude estimation, and we might ask how it stands up in comparison with actual data.

There is a widespread belief that in many cultures women are more talkative than men. This idea was tested in a research project where American university students wore a voice recorder.[13] The data showed that both male and female students spoke about 16 000 words per day. At two words per second, this corresponds to about two hours of uninterrupted talk. University students probably talk much more and much faster than, for instance, elderly people, so our estimate of 200 million words per second should be in the ballpark.

How much we talk to other people may be a measure of social competence, and so is the number of people we "know." The word "know" requires a clarification. Let it mean that when two persons meet, they both immediately remember the other person's full name. That definition excludes many acquaintances whose names we recall only with the help of, for instance, a roster. How many people do you know, with this definition? As with the example of being talkative, individual variations are large, but we may still come up with a reasonable answer. Often, the people we know can be connected with specific groups, containing typically 25–50 persons. These groups could consist of relatives, schoolmates, people at our workplace, neighbors, members of clubs or organizations, and so on. Perhaps we can identify 10 to 20 such groups. The number of people we "know" typically varies from a few hundred up to about a thousand. Interestingly, the number may reach a maximum when we are about 60 years old because memory is deteriorating and because people we have known pass away.

The number of people whose names *we* know is much larger, since we then must include celebrities and all others who are known to us but who don't know our names. Even larger is the number of people we *meet* during a lifetime. As in the case of "knowing," we need a definition of what is meant by the word "meet." Let it be quite restrictive and include only those we mutually exchange words with like the shop assistant, the bus conductor, the waiter, someone in the gym or on the beach. This is a more difficult problem than an estimation of the number of people we know. Here is a useful trick. Consider the previous week. How many *new* people did you meet, based on the definition that you talked to each other? In this case it is likely that you have met the shop assistant, the bus conductor, and so on many times before, so they should not be included in the count. Perhaps there were only a few *new* people you "met" the previous week, but of course you *saw* many more. If you are not living a very active life, the number of new people you meet in a year may be just a couple of hundred, or a thousand, even though you live in a city with one million inhabitants. In the extreme case that you have lived all your life in the same small village and do not travel much, the number of people you meet in a lifetime may be less than a few thousand, with our restrictive meaning of "meeting." Not many compared with the 7000 million people on our planet.

Hit by Returning Rocket

Would you hesitate to be in an area where rockets are tested if you were told that in the next launch the chance that someone would be hit by the returning rocket was 10^{-6}? This is not a silly question; it has a serious background, as we will now see. Esrange is the name of a rocket test site in the north of Sweden, above the Arctic Circle. Because it is located within a vast and uninhabited land, but still with a town and an airport nearby, it has become one of the world's leading centers for the launch of sounding rockets. These rockets follow the parabolic path of free flight at a very high altitude before they return to the Earth. In spite of all precautions, one cannot rule out with absolute certainty that it will not hit a person on the ground. To get permission for launch, administrators of the test site must demonstrate that the probability for such an accident is less than a certain number, n.

In a 1964 agreement between the Swedish government and the European Space Research Organization, values were assigned to n: $n = 10^{-6}$ for the public and 10^{-5} for the personnel involved in the launch. The *actual* chance for someone to be hit may be much smaller than these limit values. In the estimation of the risks, Esrange makes the amusing assumption that a person on the ground has the effective cross-sectional area of π m^2, corresponding to a radius of 1 m. This size is reasonable, since the rocket parts are rather small, and fatal only in a direct hit. There have been many launches since 1964 and no serious accidents.

How small is an overall risk of 10^{-6}? Are we almost always engaged in activities at this risk level? This is a meaningful question only after some kind of normalization of the activity. If we are flying, it may be relevant to consider the risk per flight, rather than per traveled distance, since most airplane accidents occur at takeoff or landing. On the other hand, the probability of a car accident is better related either to the time spent in the car or to the distance traveled. In the safest countries in the world (e.g., the Netherlands) there are fewer than about 0.5 automobile fatalities per year and per 100 million passenger kilometers. One hour's drive with two people in a car traveling at an average speed of 50 kilometers per hour results in 100 passenger kilometers. The fatality rate during such a ride thus is about $0.5/10^6$. In North America and most countries in Western Europe the fatality rate is somewhat higher, near 1×10^{-6}. Many people are afraid of flying but still don't hesitate to drive one hour to the airport and back again. From the estimate of car accidents we conclude that the total chance of a fatal car crash during the drive to the airport typically is approximately 10^{-6}, or the same as the highest acceptable risk level connected with a single launch at Esrange. But the latter case refers to one fatality in the "pool" of people at risk. If you enter the launch area, there are many others in the pool.[14] Further, the actual risk level at the rocket site is likely to be much smaller than what lies at the margin-for-launch permission.

Risk researchers have noted that many people seem to accept a certain total risk level as being "normal."[15] If their risks in some areas of life are reduced, they compensate for that by taking larger risks in other areas. If they purchase better brakes and tires for their cars, people may drive at higher speeds, and so on. This inclination to maintain a constant risk level is called the theory of *risk homeostasis*.

Table 1.7 gives the probabilities of fatalities associated with some

Table 1.7. Number of U.S. fatalities from various accidental causes, expressed as the population size, *N*, for which one fatality per year is expected (averages for 1999–2003)

Cause	*N*
Motor vehicles	8000
Poisoning	20 000
Drowning	80 000
Fires	90 000
Electric current	700 000
Lightning	6 000 000

Source: Data are rounded numbers selected from statistics published by the U.S. Department of Transportation Pipeline and Hazardous Materials Safety Administration (PHMSA).

activities, and normalized in different ways. Of course, these numbers can have large uncertainties and vary with the individual, but at least they give a reasonable order-of-magnitude estimation.

1.4 Estimates

There are many planets in the universe that might have intelligent life, and many more atoms in a grain of sand than there are grains in a sandpile, and there is not enough paper in the world to wrap up our planet.

Is Anybody Out There?

Arguably the most challenging estimation problem is how many planets there could be in the universe with life forms like those on Earth. A conference in 1961 at the National Radio Astronomy Observatory in Green Bank, West Virginia, was the starting point for the search for extraterrestrial intelligence, abbreviated SETI. The conference discussed the famous mathematical expression that is sometimes called the Green Bank equation but is more often referred to as the Drake equation because it was formulated by Frank Drake in preparation for the conference. The equation seeks to quantify the number, N, of civilizations in our galaxy (the Milky Way) whose radio emissions are now detectable on the Earth. It reads

$$N = R_{\text{star}} \times f_{\text{planet}} \times n_{\text{ecology}} \times f_{\text{life}} \times f_{\text{intelligent}} \times f_{\text{communication}} \times L_{\text{lifetime}},$$

where

R_{star}	=	rate of formation of stars in the Milky Way expressed as an average number per year
f_{planet}	=	fraction of those stars that form planetary systems
n_{ecology}	=	number of planets in those systems that are ecologically suitable for life forms
f_{life}	=	fraction of those planets on which life forms actually develop
$f_{\text{intelligent}}$	=	fraction of planets with life that then evolves to an intelligent form
$f_{\text{communication}}$	=	fraction of planets where the intelligent life also develops technology that releases detectable signals
L_{lifetime}	=	average "lifetime" during which such advanced civilizations release signals

The four fractions, f, in the equation must lie between 0 and 1.

As an extreme case, assume that all stars are like our solar system, with $n = 1$ and all $f = 1$. Then the equation reduces to

$$N = R_{\text{star}} \times L_{\text{lifetime}}.$$

The mean rate of star formation in the Milky Way is well understood, and $R_{\text{star}} = 10$ per year is widely accepted to be reasonable. Radio communication has existed on Earth for more than a century. Taking this fact to indicate that L_{lifetime} is at least 100 years, the Drake equation would give at least about 1000 civilizations sending out radio signals that we might detect. One may ask if the equation should not take into account that a signal cannot propagate faster than with the speed of light. If we receive a message now, the other civilization may have died out before it could receive our answer to the call. However, the propagation time does not affect the estimated N from the Drake equation because it assumes a "steady state." Civilizations that become extinct before we receive their messages have, somewhere, been replaced by other and equivalent civilizations. If technologically advanced civilizations can avoid self-annihilation for a period L_{lifetime} at least as long as the existence of *Homo sapiens* (over 100 000 years), the consequences suggested by the Drake equation are enormous for how we view ourselves. But perhaps our solar system is far from typical. Some people would claim that it is unique, in the strongest sense.

The 1961 conference suggested that $R_{\text{star}} = 10$ per year, $f_{\text{planet}} = 0.5$, $n_{\text{ecology}} = 2$, $f_{\text{life}} = 1$, $f_{\text{intelligent}} = 0.01$, $f_{\text{communication}} = 0.01$, $L_{\text{lifetime}} = 10^4$ years. These numbers give $N \approx 10$. Now let us take a closer look at the values assigned to the parameters. At the time of the Green Bank conference, no one had any information about the existence of planets around stars other than our own sun. Planets are too small to be directly observed from the Earth with a telescope. A breakthrough came in 1995, when variations in the motion of the star 51 Pegasi were interpreted as caused by an orbiting planet. Fifteen years later, about 300 stars with planets had been identified. Astronomers usually assume that life requires the presence of liquid water and an atmosphere. This limits the temperature (i.e., the distance to the star) and the minimum size of the planet (the atmosphere escapes from a small planet). If our solar system is regarded as typical, one might expect that there is often at least one planet with conditions suitable for the development of life. Therefore, there is no reason to change the original assumption $f_{\text{planet}} \times n_{\text{ecology}} \approx 1$.

The next question is whether life actually develops on planets that are "suitable" for life. It has been argued that because life was established early in the history of the Earth, and because there are also life forms that evolve under the harsh conditions of darkness and the extreme pressure at ocean depths, f_{life} should be close to 1. It is difficult to estimate the probability $f_{\text{intelligent}}$ that life, once it has been created, also develops to an intelligent form. Equally difficult to estimate is the probability $f_{communication}$ that intelligent life at some stage develops a technology that sends out signals that we can detect.

The original suggestion that civilizations may be able to communicate with radio signals for about 10 000 years was considered by many to be far too long a period. Human self-annihilation through nuclear war was being much debated at the time of the Green Bank conference, and there are of course many other scenarios in which a technologically advanced society perishes, due to actions of the inhabitants or to natural disasters. But even if the typical $L_{lifetime}$ is much shorter than 10 000 years, the Drake equation suggests that the search for extraterrestrial life is worthwhile, and many enthusiasts are engaged in that endeavor.

Attempts to answer the question of whether there is life elsewhere in the universe, or whether we are alone, are of course not limited to modern science. For instance, St. Thomas Aquinas argued at the end of the thirteenth century that we are alone. If God had created other worlds, he said, they would either be similar or dissimilar to ours. But it would not be consistent with divine wisdom to create several similar worlds. On the other hand, if there were several dissimilar worlds, each of them would not contain everything. Thus, none of them would be perfect—and a perfect God would not create imperfect worlds.

Sand, Sibyl, Olympic Medals, and Homeopathy

Are there more grains of sand in a sandpile than there are atoms in a single grain of sand? This is not a precisely formulated problem, because the size of a sandpile is not well defined. The size of a grain of sand is also somewhat uncertain. Geologists usually call rock particles "sand" when the grain diameter is 0.06–2 mm (table 1.8). A crude calculation shows that a sandpile with as many grains as there are atoms in a typical single grain is very big indeed.[16] Its volume is about 10^{10} m^3, corresponding to a cube with sides longer

Table 1.8. Classification of soil by grain diameter

	British standard	Krumbein φ scale, USA[a]	
Name	mm	mm	inch
Boulder	>200	>256	>10
Cobble	60	64	2.5
Gravel	2	2	0.16
Sand	0.06	0.0625	0.005
Silt	0.002	0.0039	0.00015
Clay	<0.002	<0.0039	<0.00015

[a] The Krumbein φ (phi) scale for grain size, sometimes used in the United States, is a logarithmic scale in which φ is a number on the scale defined as $\varphi = -\ln(D/D_0)/\ln 2$, where D is the diameter of the particle and D_0 is a reference diameter, $D_0 = 1$ mm. For a given φ we have $D = D_0/2^{\varphi}$.

than 1 km (0.6 mile). The problem thus has a definite answer, in spite of its imprecise nature.

In *Metamorphoses*, by the Roman poet Ovid, Apollo asks the Sibyl to choose whatever she wishes. She asks for as many years of life as there are grains of sand in her hand. But she refuses his offer of eternal youth in return for her eternal love, and she washes away until only her voice remains. From the example above we see that 1000 years of life corresponds to a few grams of sand, so the Sibyl lived a long life. In Buddhist texts there is an expression "the sands of the Ganges," used to represent an incalculable number of worlds in the universe.

Here is another illustration of how many atoms there are in a sample that people would consider as very, very tiny. At the Olympic Games in Beijing 2008, the Norwegian equestrian team received the bronze metal in jumping but was later disqualified when one horse tested positive for the banned substance capsaicin. This substance, which occurs naturally in various kinds of pepper, can give a burning feeling on the tongue. Capsaicin was smeared on the horse's front legs.

In 1912 the American chemist Wilbur Scoville introduced a measure of pepper strength, which now bears his name. Pure capsaicin was rated as 16 million scoville. This value is defined such that if a specimen is diluted with syrup 16 million times, a person in a test panel can no longer decide if the specimen contains capsaicin or not. But there are still very many capsaicin molecules in a test sample even at that dilute limit. The relative molecular mass of capsaicin ($C_{18}H_{27}NO_3$) is $(18 \times 12) + (27 \times 1) + 14 + (3 \times 16) = 305$, which means that 305 grams of capsaicin contain 6×10^{23} capsaicin molecules (Avogadro's constant). If we diluted such a 305-gram sample 16 million times, there would still be about 10^{14} molecules of capsaicin in a sample the size of what you can get on a teaspoon. Today we use chromatography instead of relying on the subjective results from a test panel.

In sharp contrast to the large number of atoms in a grain of sand is the lack of any "potent" substance (atoms or molecules) in homeopathic products. They are made from substances that are diluted in water or alcohol and succussed (shaken vigorously). A dilution of 1 to 10 is denoted 1X, a dilution 1 to 100 denoted 2X, and so on. Similarly, if each dilution is by a factor of 100, the products after dilution are denoted 1C, 2C, and so on. The letters X and C come from the Roman numerals X = 10 and C = 100. On the market there are remedies labelled, for instance, 28X. That means a dilution of 1 to 10^{28}. Let us assume that one had as much as 1 mole of the original substance, or as many grams as are given by its relative molecular mass. It is not likely to have a volume larger than 1 liter. Dilution 10^{28} times would be equivalent to mixing this amount with 10^{28} liters of water—or 10 000 times the volume of the Earth. (Of course, not that much water is needed. After each dilution one takes out only a very small portion to be used in the next dilution.) Since Avogadro's constant is 6×10^{23}, a homeopathic product denoted higher than 24X or 12C is likely to contain not even a single molecule of the potent substance, even if one started with as much as 1 mole. Sometimes the believers in homeopathy refer to a remaining "quantum memory"—a complete misunderstanding of quantum physics. Homeopathy is a pseudoscience, a field that pretends to rest on scientific grounds but is pure nonsense.

Cover the Earth with Paper

The Bulgarian artist Christo is known for wrapping up large objects, such as the Pont Neuf ("New Bridge") in Paris and the Reichstag (Parliament building) in Berlin. We shall attempt something much larger—wrapping up our planet. Suppose that we have access to all the paper there is in the world. Then, by some magic, we also have access to all the paper that has ever been produced, even though it might now be destroyed or recirculated. Take all this paper, and put it in a single layer on the ground. Would there be enough paper to cover the Earth's entire land area?

Since we are asked only to decide if it is possible, there is no need to worry about details in the first attack on the problem. It could be that the area of the paper is many orders of magnitude larger than the area of the Earth, or many orders of magnitude smaller. In these cases even a very crude estimate will be sufficient. But if it turns out that the two areas are of about the same order of magnitude, we must reconsider the validity of the assumptions. Estimation problems can often be approached in many, quite different ways. A good problem solver uses several of them and compares the results.

It is not very difficult to estimate the land area of the Earth, but let us look it up in an encyclopedia. The value 1.5×10^{14} m^2 assumes that the Earth's surface is essentially flat, so that even irregularities like mountains are ignored. Our problem refers to this case. For reasons soon to be clear, we also factorize the total area as

$$1.5 \times 10^{14}\ \text{m}^2 = (5\ \text{m}^2) \times 400 \times 75 \times 10^9.$$

Next we turn to the paper. It comes from newspapers, envelopes, wrapping paper, toilet paper, books, and the entire miscellany of paper products. A common mistake in many estimations is that of double counting—that is, the same contribution to the total is included two, or several, times. A useful trick for our problem is to consider how much paper is associated with a single person (for instance, you), and then scale that quantity up. For each person we therefore consider only the paper that she or he is *the last person* to handle for something personally useful, before the paper goes to recycling, is washed down the sewer, is burned in a power plant, or is destroyed in some other way.

Now think of a typical day in your life, and the paper that was "used" for the last time that day. In the industrialized countries, newspapers make up a large part. A thick tabloid-sized newspaper with about 40 pages (20 sheets of paper) has an area of about 5 m^2. Papers like the *New York Times* and the *Frankfurter Allgemeine Zeitung* have a larger page size and are also much thicker on certain days, but at this stage we will not be concerned with such details.

According to the factorization shown above, we would need more than 10^9 thick newspapers per day (400 papers per year) during 75 years in order to cover the Earth. This seems to be an overestimation of the actual situation, suggesting that newsprint alone is insufficient to cover the land area of the Earth, although it would cover quite a bit of it. That forces us to take a closer look at the problem. The area of a single newspaper is often less than 5 m^2. The consumption of paper has increased rapidly during the last century, and perhaps we can ignore all newsprint produced more than 75 years ago. The population of the countries where large daily newspapers are widespread is about 10^9 (equal to the population of Europe and North America), but usually there is at most one paper per household, and not one per person. Nevertheless, the area of paper for newsprint can cover an appreciable fraction of the Earth's area, and we must look for a completely different estimation approach, one that will require access to more data.

In recent years, the world production of newsprint has been about 4×10^{10} kg. The mass per area of such paper is about 40 g/m^2, giving a total area of 10^{13} m^2. To cover the land area of the Earth therefore requires the present annual production to continue for about 150 years. What about other kinds of paper? The world's total annual paper production is about 4×10^{11} kg. If all of it goes to paper for copying machines (80 g/m^2), we would need about 40 times the present annual paper production to cover 1.5×10^{14} m^2.

Although our estimate has some uncertainties, we can conclude that all the paper that has ever been produced is just about enough to cover the Earth's land area. It would not suffice, however, to wrap up the entire Earth.

2
Measures

2.1 What Is It on a Scale?

Why one step on the logarithmic Richter scale increases energy by a factor of 32 and not by 10, why the Three Mile Island nuclear accident is not considered serious, and on whether we should worry about being hit by a meteorite.

The Richter Scale

In 1902 the Italian seismologist Giuseppe Mercalli devised a scale to classify earthquakes. It had 12 steps, with each step described by the observed effects or the effects felt by people. As an example, taken from the revision of the scale in 1931, level VIII is described as follows:

> Damage slight in specially designed structures; considerable in ordinary substantial buildings, with partial collapse; great in poorly built structures. Panel walls thrown out of frame structures. Fall of chimneys, factory stacks, columns, monuments, walls. Heavy furniture overturned. Sand and mud ejected in small amounts. Changes in well water. Persons driving motorcars disturbed.[1]

The structure of this scale is very similar to the well-known Beaufort scale devised by Britain's Admiral Frances Beaufort (1774–1857). Beaufort's scale went from 0 (no wind) to 12 (hurricane). Level 8 on the scale (gale; 34–40 knots) is described as follows: "Breaks twigs off trees and impedes walking against the wind. At sea foam is blown in well-marked streaks."[2]

The scales of Mercalli and Beaufort have the appeal of describing in simple terms what is going on, but they are not very precise and rely on the judgment of individuals. In the 1930s, the instruments used to register earthquakes were significantly improved. Charles F. Richter was then working with seismology at the California Institute of Technology in Pasadena.[3] He had just obtained his Ph.D. in theoretical physics for work in the new field of quantum mechanics. Unable to find suitable employment as a theoretical physicist, he took on a part-time job analyzing seismic data. Journalists frequently contacted the seismologists and wanted to know the size of the earthquake. Richter devised a scale that yielded a single number and could be used to answer such questions.

Richter's scale was built primarily on the maximum amplitude, in millimeters, registered on his seismograph. But the amplitude varied, of course, with the distance to the earthquake. Therefore, Richter normalized the data so that they corresponded to the amplitude of a disturbance created 100 km away. That disturbance propagates as two waves (called S and P) with known speeds. From the difference in arrival time to the seismograph one obtains the distance. Based on these two parameters (amplitude in millimeters and time difference in seconds), Richter calculated a dimensionless quantity called the magnitude, M_L, of the earthquake (the subscript L means "local"). It could easily be read from a graph such as the one in figure 2.1 by drawing a straight line connecting the data for amplitude and time. Such graphs, called nomograms, were frequently used for rapid calculations at a time when an alternative was to use a slide rule.

The magnitude varies logarithmically with the amplitude, A, registered by the seismograph. The total energy released in the earthquake, however, varies approximately as $A^{3/2}$. An increase of the magnitude by one unit therefore means that the energy increases by approximately a factor of $10^{3/2} \approx 31.6$.

The Richter scale satisfied the needs of journalists, but it did not capture the information of relevance to the seismologists. For instance, it was intended primarily to describe earthquakes in southern California in the range $M_L = 3-7$. A different measure, M_w, called the moment magnitude, was later introduced, and several

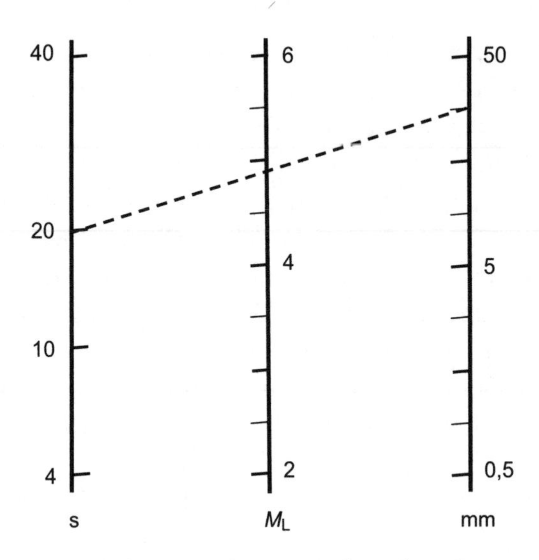

Fig. 2.1. A nomogram that gives the magnitude M_L from the seismograph's recorded amplitude s (in millimeters), and difference in arrival time t (in seconds) between the S and P waves as shown by the dashed line

other measures have also been constructed.[4] The Richter magnitude and the moment magnitude of a given earthquake are only approximately equal, but the traditional magnitude on the Richter scale is so deeply rooted in people's minds that it is always used by the media, in spite of its scientific shortcomings.

The story of the Richter scale, its predecessors, and its subsequent replacement by other measures of earthquake strength is a good illustration of how science takes step after step to improve our understanding and description of the complex real world. Knowledge is only provisional!

Nuclear Incidents and Accidents

People will forever associate Chernobyl and Three Mile Island with nuclear power plant disasters. Or at least they will do so for a very long time to come if we, in the spirit of this book, are cautious in our choice of words. Like *forever*, the word *disaster* is also vague and is

Table 2.1. The International Nuclear Event Scale (INES)

Scale no.	*Description*
0	No safety significance
Incident	
1	Anomaly
2	Incident
3	Serious incident
Accident	
4	Accident with only local consequences
5	Accident with local consequences but a risk for environmental consequences
6	Serious accident
7	Major accident

often used carelessly about things in our lives that are far from really serious. In order to put incidents and accidents in nuclear power plants into perspective, the International Atomic Energy Agency (IAEA), in cooperation with the Organisation for Economic Co-operation and Development (OECD), formulated the International Nuclear Event Scale (INES), shown in table 2.1. This scale applies also to the transport, storage, and use of radioactive material and radiation sources, including, for instance, radiography and use in hospitals, but does not apply to military applications. The scale ranges from 1 to 7; levels 1–3 are called *incidents* and levels 4–7 *accidents*. Events that have no safety significance are denoted 0 and lie outside of the scale.

Since the numbers in the INES scale cannot be unambiguously related to measured quantities, they must be explained in words. The descriptions are based on criteria that refer to three different areas:[5]

- Consequences for people and the environment
- Local consequences
- Reduction of safety barriers

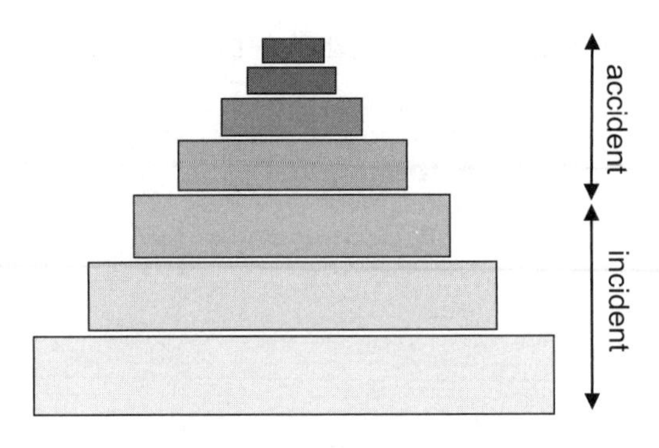

Fig. 2.2. Graphic illustration of the seven levels of the INES scale

The first two categories deal with the *actual* consequences of what has happened, and the third category deals with *potential* consequences.

The Chernobyl accident in 1986 was characterized as 7 on the INES scale because of its severe and widespread effects on people and environment. The Three Mile Island accident in 1979 was characterized as level 5 because of the severe damage to the reactor core. Three Mile Island falls into the category "local consequences." An example of a so-called serious incident (level 3) is an instance when a package containing radioactive iridium was sent without proper protective cover, and personnel handling it received an increased radiation dose. The loss or theft of radioactive sources may also be classified as level 3. The discovery of a malfunctioning emergency electric power supply in a nuclear power plant meant a reduction in safety barriers and was classified as level 2 on the INES scale. Examples of incidents at level 1 are the loss of a moisture-density gauge and a temporary exceedance of the operating limits for power output. Such anomalies may occur every year in a commercial nuclear power plant, while the frequency of incidents at level 3 is much less than one per year. The INES scale is an important tool in nuclear safety work. It is universally adopted, and incidents or accidents characterized as level 2 or above are promptly reported to the IAEA.

The INES levels are often illustrated graphically as shown in figure 2.2. However, such a picture may unintentionally give the impression that each step represents a rather small change. In fact, one step corresponds to an increase in seriousness by at least a factor of 10. The official words used to characterize the levels may also cause communication problems about risks. If the public were asked whether they consider the Three Mile Island case a major accident, or at least a serious accident, almost everyone in the United States would say yes. But the phrase "major accident" refers to level 7, and "serious accident" to level 6. The Three Mile Island accident was classified as level 5, meaning an accident with local consequences but with a risk of environmental consequences.

Natural Threats

Depending on where we live, natural threats may be an important aspect of our daily lives. Natural threats could be events that occur more or less regularly—like hurricanes, flooding, wildfires, and avalanches—or rare events like volcanic eruptions. Meteor impacts are extremely rare but can have devastating consequences. The last example highlights the two important aspects of the concept of risk: the probability that something will happen and the consequences if the event does occur. Often these two considerations are weighted together, as is done in the calculation of insurance rates.

When natural threats are classified according to some scale, it can be either the probability or the consequence that one wants to quantify. Avalanche and wildfire warnings represent the first category. The public should be alerted so that they take precautions, although nothing may happen. Volcanic eruptions represent the other extreme: there can be no meaningful measure of the probability, but it is of interest afterwards to have some measure of the strength of the eruption. In many cases the threats and consequences are so complex that they can only be described in qualitative terms. Then it is very common to define a scale through rubrics—for instance, with five levels. Level 1 may refer to a very small risk or very small

Table 2.2. Classification of avalanche danger by Colorado Department of Natural Resources

	Avalanche probability	
Danger	*Natural*	*Human-triggered*
Low	Very unlikely	Unlikely
Moderate	Unlikely	Possible
Considerable	Possible	Probable
High	Likely	Likely
Extreme	Widespread avalanches certain	Widespread avalanches certain

consequences, or it can be a measure of the least dangerous event of its kind.

Unlike the INES scale for nuclear events, many of the scales describing natural events are not globally accepted, but are developed and used in certain geographic areas. Of course, they are likely to be similar if they describe the same phenomena, but a direct comparison is often not possible. Further, different words may be used. For instance, what is called a hurricane in the Atlantic Ocean is called a super-cyclonic storm in the North Indian Ocean, a severe tropical cyclone in Australia and the southwestern Pacific, and a super-typhoon in the northwestern Pacific.

The classification of avalanche danger in table 2.2 is typical of scales that are meant to advise the public of natural threats. In this case, travel (or, e.g., skiing) is generally safe at the lowest risk level, and normal caution is sufficient. At the most dangerous level, travel (or sports) should be confined to low-angle terrain well away from path run-outs.

The levels of the various hurricane scales are based on wind speed, usually measured as 10-minute averages. Thus, there is an objective basis for the assigned level. For other phenomena, like avalanches and wildfires, the situation is so complex that no unambiguous measure can be defined. Nevertheless, authorities can rely on algorithms

Table 2.3. Examples of levels in the Torino scale for asteroid threats (abbreviated descriptions)

Threat level	*Description*
3	A close encounter, meriting attention by astronomers, with a 1 % or greater chance of a collision capable of causing regional devastation.
7	A very close encounter by a large object, which poses an unprecedented but still uncertain threat of a global catastrophe.
10	A collision is certain, capable of causing global climatic catastrophe that may threaten the future of civilization as we know it now, whether impacting land or ocean. Such events occur on average once per 100 000 years, or less often.

that use temperature, moisture, and other measurable quantities as input data to produce a numerical output that determines the assigned risk level.

Not all scales go from 1 (or 0) to 5. The strength of a volcanic eruption is given by the *volcanic explosivity index* (VEI), on a scale from 0 to 8. The assigned value is an estimate that is based on the amount of lava, ash, and other material in the explosion, and also on the altitude in the atmosphere reached by that material. For instance, VEI = 2 corresponds to about 10^6 m^3 of material reaching 1–5 km. Such eruptions occur about once a week on our planet. The Mount St. Helens eruption in 1980 had a VEI of 5, and the Krakatoa eruption in 1883 had a VEI of 6. The Yellowstone eruption about 640 000 years ago, and the Toba eruption 73 000 years ago, both reached the highest value, VEI = 8.

Asteroid threats are categorized by the *Torino scale* (adopted in the Italian city Torino, or Turin, during a workshop in 1999 and revised in 2005). The Torino scale is designed to indicate the risk of collision between an asteroid and the Earth, and the possible consequences. The scale has levels from 0 to 10. Some of them are defined as in table 2.3.

2.2 Comparing Apples and Oranges

On whether the quality of life is highest in the Nordic countries, whether it was unfair at the 1932 Olympics in Los Angeles when the decathlon gold medal went to the United States and not to Finland, and a clever way to award points in the scout camp competition.

Human Well-Being and Poverty

The Pakistani economist Mahbub ul Haq and several others have devised a measure of human well-being in a country, called the *Human Development Index* (HDI). Since 1990 it has been used in the annual report from the United Nations Development Programme. HDI is formed as an arithmetic average of three indices relating to life expectancy, education, and economy. The index for each of these areas (called dimensions) has the general normalized form

$$\frac{\text{actual value} - \text{minimum value}}{\text{maximum value} - \text{minimum value}}.$$

The maximum and minimum values refer to the largest and the smallest number, within the considered dimension, for any country. Thus, for each of the three indices there will be one or several countries that are given the value 1, or the value 0. A country that reaches the maximum value in each of the three dimensions would get HDI = 1, and a country that scores 0 in all three dimensions would get HDI = 0. In practice, HDI ranges from about 0.3 to about 0.97. The world average is about 0.75. Values below 0.5, obtained for many countries in Africa, are considered as representing low development. The rankings of individual countries vary somewhat over the years, but the five Nordic countries (Denmark, Norway, Sweden, Iceland, and Finland) have consistently been in the top group, along with Australia, Canada, the United States, Japan, and several European countries.

In order to calculate HDI we need explicit formulas containing the raw data in the three dimensions. The life expectancy measure is defined as $(p - 25)/(85 - 25)$, where p is the life expectancy at birth, in years. The largest and the smallest value of this measure may change with time, but after normalization (as in the displayed formula above), the resulting life expectancy index always lies between 0 and 1.

The education index is defined as 2/3 times an adult literacy measure plus 1/3 of a combined primary, secondary, and tertiary enrollment measure. Both of them are first normalized as above, and then normalized again after the 2/3 and 1/3 weighting, so that the final index lies between 0 and 1. The economy index is defined as $(\log_{10}N - \log_{10}100)/(\log_{10}40\,000 - \log_{10}100)$, where N is the gross domestic product (GDP) in US dollars. In 2008 the UN Development Programme issued a new report based on improved GDP data derived from purchasing power parities (PPPs).

Not surprisingly, HDI has been criticized. One argument has been that it measures average achievement rather than deprivation. That has led to the definition of a *human poverty index* for developing countries, HPI-1. Like HDI it is based on three indicators, but these are expressed in percent and are therefore already normalized. The first indicator is the probability of *not* surviving to age 40 (which is then multiplied by 100), the second indicator is the adult illiteracy rate (percent), and the third indicator is the sum of half the percent of the population not using an improved water source and half the percent of children who are underweight-for-age. The three indicators are then combined to yield HPI-1 through a mathematical expression.

There is also another human poverty index, HPI-2. It has four indicators, measuring the probability of not surviving to age 60, literacy, income below poverty line, and rate of long-term unemployment.[6] Finally, there is a gender-related development index, GDI, which adjusts the human development index to reflect the inequalities between men and women. The procedure is similar to that of the average HDI, but the indices calculated in each dimension are combined to a so-called equality distributed index through an elaborate formula that penalizes differences in achievement between men and women.[7]

Track and Field

The seven events in the women's heptathlon are the 100-m hurdles, the high jump, the shot put, and the 200-m sprint on the first day, followed the second day by the long jump, the javelin throw, and the 800-m race. In the men's decathlon the events on day one are the 100-m sprint, the long jump, the shot put, the high jump, and 400-m, followed on day two by the 110-m hurdles, the discus throw, the pole vault, the javelin throw, and the 1500-m race. The results from each event are converted to numbers (points), which are then added with equal weight. An athlete who is very good but still below the world elite in the particular event will achieve about 1000 points in each event.

The simplest approach to a scale of points would be to use a linear relation, so that, for instance, an improvement by 10 cm in the men's high jump from 1.90 m to 2.00 m would be worth as much as an improvement from 2.30 m to 2.40 m. Similarly, an improvement by 0.2 s in the women's 100-m race from 13.3 s to 13.1 s would be worth as much as going from 11.5 s to 11.3 s. Use of such a linear scale, however, makes it much easier for mediocre athletes than for the elite athletes to improve their score—an undesired effect. Scoring tables are therefore progressive, meaning that the same improvement in centimeters or seconds leads to a greater increase in the score at the top end of the results.[8]

The 1985 IAAF (International Association of Athletics Federations) scoring table converted a result, R, to points, P, according the formula

$$P = A(R - B)^C.$$

The parameters A, B, and C vary with the event. R is the numerical value of the performance, measured in the same unit as B. For track events, $R - B$ is replaced by $B - R$, to get a positive number. The final number of points is given by the integer value of P calculated as above. Table 2.4 gives the 1985 values of A, B, and C for the men's

Table 2.4. Parameters in the formulas of the 1985 IAAF scale for men's decathlon

Event	*A*	*B*	*C*	*Unit for R, B*
100 m	25.4347	18	1.81	s
400 m	1.53775	82	1.81	s
1500 m	0.03768	480	1.85	s
110-m hurdles	5.74352	28.5	1.92	s
Long jump	0.14354	220	1.40	cm
High jump	0.8465	75	1.42	cm
Pole vault	0.2797	100	1.35	cm
Shot put	51.39	1.5	1.05	m
Discus throw	12.91	4	1.10	m
Javelin throw	10.14	7	1.08	m

decathlon. Somewhat different values of *A*, *B*, and *C* applied to the women's heptathlon.

The world records in the 10 events in the decathlon would give points ranging from just over 1100 (110-m hurdles) to almost 1400 (discus throw), with an average of about 1250. The world record in the decathlon is an average of about 900 points per event. In the Olympic Games of 1932, which used a decathlon scale from 1912, James Busch (US) won the gold medal with 8462 points, and Akilles Järvinen (Finland) got the silver medal with 8292 points. However, if the 1985 IAAF table had been used, Järvinen would have been the winner, with 6879 points, while Busch would have come in second with 6735 points.

The decathlon scales have been changed several times — in 1936, 1952, 1962, and 1985. The change made in 1985 scaled down the advantage of performing very well in a few of the events. In 2008, after a careful evaluation of progress in recent years, the IAAF published new scoring tables for comparing performances in different events. About half of the scores in the 1985 tables were changed, most of them in walking and long-distance running, but not the events listed in table 2.4.

At Scout Camp

Humans seem to have an innate desire for competitions, between individuals or teams. Cheating and unfair play is not accepted—not even in a friendly family game. In some games, pure chance plays a large or completely decisive role, but in serious competitions we try to have rules such that the winner will also be "the best." The scales devised for the heptathlon and the decathlon, discussed in the preceding section, represent that effort taken to the extreme. Competitions between teams at a Scout camp or between schools in a local school district are often said to be for fun, but that may not be how the contestants see these games, particularly if the winner also gets a significant prize. When the competition means that achievements in different events are aggregated to a final score, one should strike a balance between simplicity and fairness, and that is what we will now discuss.

Thinking like scientists, we can grossly simplify the problem to make the point. Assume that there are four Scout teams in a competition with only two events: woodcraft and tug-of-war. The woodcraft event has 20 questions about the natural environment. Teams receive one point for each correct answer—for instance, the name of an edible plant or the identification of an animal dropping. Of course, we do not want to make it so difficult that most teams score very poorly. On the other hand the questions should not be so simple that the teams get almost all of the questions right. As a compromise we might expect the teams to score in the range of 8 to 16 points each.

The tug-of-war event is of a very different nature. If all four teams meet once, it is likely that one team will get three victories, and the other three teams two, one, and none, respectively. How should points be awarded for this event so that it has the same weight as woodcraft? Since woodcraft has a maximum of 20 points, we could decide that the winner in tug-of-war gets 20 points, and the other teams get 15, 10, and 5 points, respectively. But then the spread between the best and worst teams would be 15 points, while the expected spread in woodcraft was only 8 points. In practice, tug-of-war is then given double the weight of woodcraft.

We could try to make the two events equally important by letting the tug-of-war points be 20, 17½, 15, and 12½, respectively. Then the difference between the first and the last team is 7½ points. But we could also achieve the same goal with points of 7½, 5, 2½, and 0 for the tug-of-war. The two scales look very different, but they give the same ranking of the four teams after combination with the woodcraft event. Nevertheless, many people would object to these alternative scales—the one with points from 20 to 12½ as well as the one with points from 7½ to 0—because they appear incompatible with the scale for woodcraft where the points theoretically range from 20 to 0.

Our example highlights that in terms of fairness, the important feature is the expected *spread* in points between the competitors, from the top to the bottom—not the maximum number of points that can be awarded in the event. We can illustrate this in another example. Suppose that nine teams compete in many very different events. The result in one event may be expressed as a time and in others as a length, the number of correctly answered questions, or the number of points awarded through the judgment of a jury. We seek a method to aggregate these achievements in such a way that all events are given equal weight.

A natural solution, and also the simplest one, would be to rank the competitors from 1 (bottom) to 9 (top) within each event and give them points equal to their ranking number. That would guarantee an equal spread in each event. In this case we might also consider giving the winning team in each event 100 points, the next team 90, and so on, down to 20 points for the ninth team. That might look nicer on the scoreboard and would not change anything in the outcome of the entire competition. However, the procedure with points based directly on the ranking number ignores the fact

Table 2.5. Two scales for points in an event with nine competitors

Simple ranking	20	30	40	50	60	70	80	90	100
Modified points	20	36	48	56	60	64	72	84	100

that the primary data recorded for a certain event (for instance, times or lengths) tend to have a Gaussian-like (i.e., bell-shaped) distribution, with many almost equal results and a few outliers on each side. Teams near the average may get quite different ranking points even though their achievements are approximately the same.

This problem can be lessened as follows. We first rank the teams on the basis of their primary result and give them points from 100 to 20 depending on their position in the ranking list. In the next step, these points are replaced by the numbers shown in the second row in table 2.5. For instance, the third team in the event might get 72 points instead of 80. The two scales coincide in the middle and at the ends, and the total number of distributed points (540) is unchanged. In the new scale, the increase in each step varies as 16, 12, 8, 4, 4, 8, 12, 16, rather than being constant. Of course, the modified scale has some arbitrariness, but it is probably fairer than a simple ranking.[9] On the other hand, we may run into communication problems if such a scale is used, since it looks awkward to those who are not familiar with Gaussian distributions.

2.3 Units

One thing the United Sates has in common only with Liberia and Burma, the horsepower output of a man, and a simple mistake with units that cost $125 million

Going Metric—Inch by Inch

Only three countries in the world have not yet adopted the metric system—the United States, Liberia, and Burma (Myanmar). But humans are slow to accept changes. The road signs in Northern Ireland, which is part of the United Kingdom, still give distances in

Fig. 2.3. The meter prototype introduced in 1889

miles. As soon as one crosses the border to the Republic of Ireland, however, the distances are given in kilometers. The same thing is seen when one drives from the United States into Canada. The weight of a person (or, to be formally correct, the mass) is given in kilograms in the metric system (*Système International d'Unités*, SI), but many older British people still think in terms of stones. One stone is 14 lb or 6.35 kg.

The definitions of the meter and the inch have an interesting history. When the meter was introduced at the end of the eighteenth century, in the wake of the French Revolution, it was fixed to be one ten-millionth of the distance between the pole and the equator of the Earth. On the basis of this definition and geodetic measurements, the famous meter bar was constructed as the prototype and kept in the archives of the French Republic (hence its name, archive meter). It was replaced in 1889 by a new meter bar, which is about 102 cm long and made of a platinum-iridium alloy with an X-shaped cross section and marks 1 m apart (fig. 2.3). The definition of the meter was changed again in 1960 to be 1 650 763.73 times the wavelength of a spectral line in the isotope krypton-86. But a spectral line is not absolutely sharp, so even this value has some uncertainty. In 1983 scientists took the radical step of defining the meter in terms of the speed of light in vacuum. Thus, the meter is now defined as *the length of the path travelled by light in vacuum during a time interval of 1/299792458 of a second.*

The progress in engineering and technology at the end of the nineteenth century meant increased demands for accuracy in the manufacture of parts. The metric system had conquered much of the world, and in 1893 the metric prototypes for the meter and the kilogram were declared fundamental standards of length and mass in the United States. But more than a century later, Americans still use inch, foot, and yard.

A big step forward in practical measuring techniques was taken in 1901 when the Swedish inventor Carl Edvard Johansson got his first patent for a set of gauge blocks made of steel. The gauge blocks were introduced in the United States by the Cadillac Automobile Company around 1908 and later adopted by Ford. The blocks came in series with different widths. By combining blocks, one could achieve an accuracy of ±0.001 mm, from 1 to 100 mm. The gauge blocks were used to calibrate the measurement tools for the automotive industry to an unprecedented accuracy. However, prototypes for the yard in Britain and the United States differed in length. It turned out that 1 inch = 25.39998 mm according to the British yard prototype, while the American yard prototype had 1 inch = 25.40005 mm. Johansson did not want to make different sets of gauge blocks for the British and the American markets, so he used the conversion 1 inch = 25.4 mm, exactly. After World War II, the definition of the inch was officially changed to exactly 0.254 m.

Horsepower and Manpower

The following dialogue was recently heard. A man remarked, "My new car has a 150-kilowatt engine." His friend looked surprised and said, "Oh, you bought an electric car?" Many people still use horsepower in connection with cars and associate watts with electricity. But the watt (W) is the common unit for all kinds of power—be it mechanical, electrical, or thermal.

The British engineer James Watt invented steam engines, which were used to pump water out of mine shafts. He wanted a measure of how much water they could pump up per hour from a certain depth. This was at a time when there was no real understanding of the concept of energy and the different forms it could take. Before the introduction of steam engines, horses were a major source of energy. Therefore, James Watt compared his engines with what could be achieved with pumps driven by horses. Because Britain used the traditional units of pound for mass and foot for distance, one horsepower was defined as the power required to lift one pound 550 feet in one second (or twice as much raised to half the height in the same time, etc.). In modern units that is approximately 746 W. The original definition of horsepower, often called mechanical horsepower, is rather arbitrary, and there are different versions of its history. We may also note that 746 W is an overestimation of the power a horse can deliver for an extended time.

While Watt's horsepower was expressed in terms of pounds and feet, the metric system, which was later introduced in Germany and many other countries, led to another definition. Horsepower was still defined by the mass that could be raised a certain distance in one second. This alternative definition specified the raising of a body of 75 kilograms one meter in one second, under the influence of the Earth's gravitational force. This unit, sometimes called a metric horsepower, is equal to 735.499 W. Another obsolete notation is 75 kpm/s, which refers to 75 kilopond meters per second. To add further to the confusion, both Watt's mechanical horsepower and the metric horsepower are denoted hp in English texts, but another symbol is often used in other languages, reflecting their word for "horse." For instance, in Germany the abbreviation is PS (for *Pferdestärke*) and in France, CV (*cheval vapeur*) or ch (*chevaux*).

The human body can be viewed as a kind of machine that takes in fuel (food and oxygen) and converts it to other forms of energy. Like all machines, our bodies are not 100 % efficient in that conversion. Roughly 25 % of the energy can be converted to mechanical work—for instance, climbing a mountain or lifting boxes from the floor to a shelf. The rest, about 75 %, becomes heat. During hard work, so much heat is produced in this inefficient conversion of food energy that we sweat heavily, which helps the body to maintain its normal temperature. Table 2.6 gives typical values for the "useful" human power (i.e., excluding the amount expended in heat production) during some activities.

Industrialized society uses vast amounts of energy, for heating,

Table 2.6. Typical human power output during selected activities

	Power, W	
Activity	*Mechanical output*	*Total output (25% efficiency)*
Sleeping	–	80
Mechanical work, moderate effort	50–75	200–300
Sports activity by elite athlete	250–350	1000–1400

transport, lighting and so on. We could estimate its magnitude with the following approach. Think of how much money you spend on energy, as an average per person and year. You must include not only your individual energy bills but also your share of what a modern society spends on, for instance, light and heat in workplaces, transport other than by private car, energy to produce goods of all kinds, and so on. It may amount to $5000 annually. Then we need a price for the energy. Very roughly, all kinds of commercial energy, known as energyware (electricity, gas, oil), cost the same per energy unit—for instance, per kilowatt-hour (kWh). If that were not the case, we could not consider using either gasoline or electricity to run cars.

For the sake of the argument, let us assume that 1 kWh of energy costs between $0.10 and $0.20. Then $5000 can buy 25 000–50 000 kWh of energy per year. A person who works, say, 2000 hours per year and produces 100 watts (that is rather heavy work) generates 2000×0.1 kWh = 200 kWh in a year. You would need 125–250 such persons to obtain the energy used by a typical citizen in an industrialized country. Our calculation is crude, but not completely unrealistic (cf. data in table 1.6 for EU countries), and it gives a perspective on our modern society. Not long ago we had to rely mainly on the energy we could deliver ourselves, plus perhaps that of a single horse.

The Loss of a Spacecraft

In September 1999, NASA (the US National Aeronautics and Space Administration) announced that the Mars Climate Orbiter had burned up in the Martian atmosphere. The spacecraft had cost $125 million. The disaster was not the result of a mechanical or electronic failure but was due to a trivial error in the use of units. NASA had specified that SI units be used, but the prime contractor had used traditional US units. For instance, the thrust of the engines was calculated in pound-seconds (lb·s) instead of the required newton-seconds (N·s).

The task of the Mars Climate Orbiter was to circle Mars at an altitude of 140 km for two years and take weather data similar to what is done from satellites around the Earth. Instead, the orbiter came as close as 60 km from Mars' surface, where the atmosphere was dense enough to rapidly destroy the spacecraft.

The subsequent investigation showed that the confusion about units was not the only reason for the disaster. The spacecraft had been on its journey to Mars for nine months. Engineers at the Jet Propulsion Laboratory in Pasadena found at an early stage that the path was not quite what they had expected. Their concerns were ignored when brought to higher levels in the administration hierarchy. Lack of communication between different organizational levels, and understaffing, were later identified as important additional reasons for the failure of the mission. These findings are in line with the conclusions of modern risk analysis. There will always be human mistakes, and it is too easy to blame an individual person for actions at the end of a chain of events. Instead, the responsibility may be at the management level. It is there that one must make proactive decisions so that the number of mistakes is reduced and, if they do occur, there that the consequences can be handled.

At the time of the loss of the Mars Climate Orbiter another spacecraft, the Mars Polar Lander, was on its way and due to arrive a few months later. It was also lost when it violently crashed onto the Martian surface instead of making a controlled touchdown. The cause was not an error in units but a technical failure. The descent engines were designed to shut down at an altitude of 3 m above the surface. NASA concluded that the spacecraft's onboard system confused the jolt from the deployment of a landing leg with ground contact and shut down the engines at an altitude of 40 m. The Polar Lander crashed at a speed of 22 m/s. (It should be said that there have been several successful missions to Mars, starting as early as 1975 with Viking 1 and Viking 2, which landed on Mars and operated until 1982 and 1980, respectively.)

There are many more examples where confusion of units has led to complete failure or near-disaster. On July 23, 1983, flight AC7067 from Montreal to Edmonton in Canada made an emergency landing because it ran completely out of fuel at a height of 41 000 ft (12 000 m). The pilots managed to glide to the abandoned Gimli airport in Manitoba and land there with no fatalities. This was a remarkable achievement by the pilots. The aircraft lost all electricity when both engines stopped. For such emergencies the aircraft had a turbine that was set in rotation by the airflow, thus providing a minimum of power.

The subsequent investigation showed that there was a long chain of circumstances leading to the near-disaster. The captain had correctly determined the fuel requirement from Montreal to Edmonton to be 22 300 kilograms. Then, to calculate how much fuel should be added, one had to convert between mass in kilograms and volume in liters. Because the fuel gauge of the aircraft was out of order, the amount of fuel was obtained by dipstick measurements. One liter of jet fuel has a mass of 0.803 kilogram. However, the ground crew was used to imperial units, in which one liter of fuel has the mass 1.77 pounds. Canada had just introduced metric units, and not all aircraft changed to that system at the same time. When the dipstick indicated that there was already 7682 liters of fuel, this was converted as $7682 \times 1.77 = 13\,597$ and reported as the amount in kilograms, although it was actually in pounds. The same mistake was made when the required additional amount of fuel was calculated. As a result, the aircraft took off with 22 300 *pounds* of fuel instead of the required 22 300 *kilograms*. Because of the problem with the automatic fuel indicator, the captain double-checked the arithmetic of the calculations but still used the wrong conversion factor. It was a tough lesson for many, not least for the management, which was accused of negligence of safety routines.

2.4 On the Road

The relatively few fatalities when Sweden changed from left-hand to right-hand traffic, the value of a human life, and why the octane rating of regular gasoline is 87 in the United States and Canada, but 95 in Europe.

Left-Hand Traffic

Most countries in the world have right-hand traffic, but there are several exceptions. In Europe, vehicles drive on the left side of the road in the United Kingdom, Ireland, Cyprus, and Malta. This is also the case in, for instance, Australia, New Zealand, South Africa, and Japan.

Sweden changed from left-hand to right-hand traffic at 5:00 a.m. on Sunday, September 3, 1967. There had been a consultative referendum on making the switch 10 years before, in 1957. Then, only 15.5 % voted for a change, out of a voter turnout of 53.2 %. In spite of such strong opposition, the Swedish Parliament decided six years later that Sweden would change to right-hand driving. The public feared that there would be a large number of fatalities. So, what actually happened? The number of accidents and fatalities fell sharply during the year after the change!

There were many possible explanations. One was that people did not drive as much as usual during that period. Obviously, to determine whether this was true we should not look just at the number of fatalities but normalize it—for instance, as the number of fatalities per one million passenger kilometers. Such statistical data were not available. But we do know the total volume of fuel for cars that was sold per year. Normalizing the fatalities as the number of fatalities per sold volume of fuel would give a good measure. That is shown in figure 2.4. The number of fatalities went down significantly, even

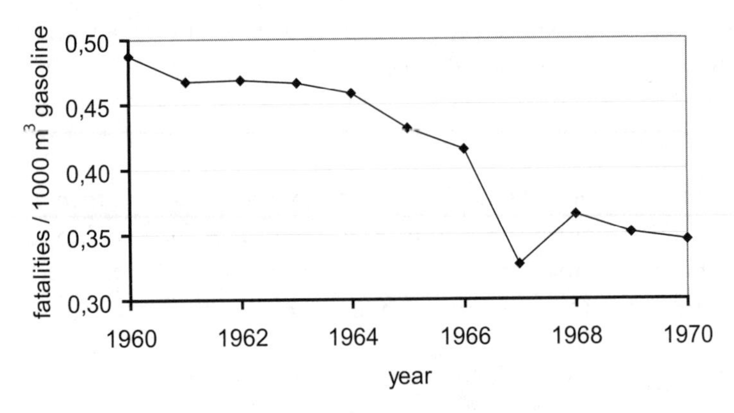

Fig. 2.4. Swedish traffic fatalities, 1960–1970. Sweden changed to right-hand traffic in 1967. The vertical axis gives the number of fatalities per 1000 m^3 of gasoline sold.

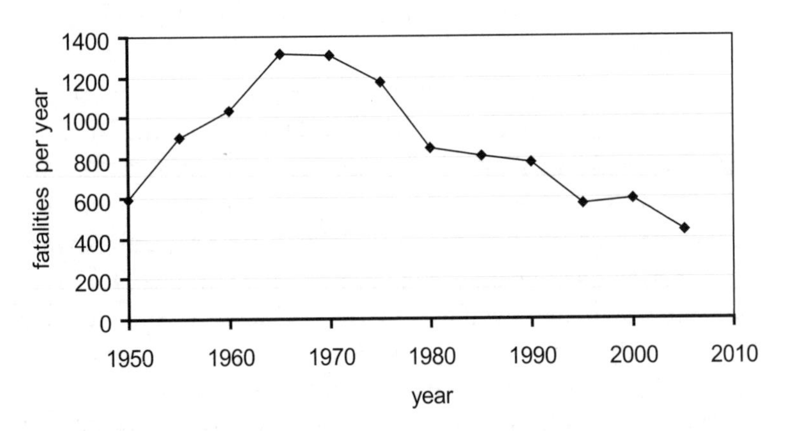

Fig. 2.5. Swedish traffic fatalities after 1950

after one allows for the decreased driving. But a year later the death rate was back to the level one would expect if there had been no change. Just after the change, people drove slowly and were more careful. There was also a massive media campaign about safe driving, before and after the change.

The overall change after World War II in the number of annual automobile fatalities per million inhabitants in Sweden (fig. 2.5) shows the same pattern as in many other European countries. First, there was an increase because with a growing economy people could afford to buy a car. In parallel with this, many measures were taken to reduce the number of accidents and their consequences. Improved roads, safer cars, mandatory use of seat belts, and so on led to a steady decrease in the number of fatalities. Many countries in the world are still in the first phase, where traffic accidents are a very common cause of premature death. For instance, worldwide there are about 200 automobile fatalities annually per one million inhabitants, as compared with about 50 in the countries with the safest driving, in spite of the much larger number of cars per one million inhabitants in the latter case.

The Value of a Life

We as individuals, and also politicians and civil servants, often put a value on a human life—indirectly and therefore usually without being aware of it. In this instance it is not a question of the lives of Mrs. Smith or Mr. Johnson or our relatives. Instead, it is a so-called *statistical life*. Suppose that certain measures regarding road safety in an area—for instance, better clearing of snow during the winter—are judged to decrease the average number of fatalities due to snow-related car accidents from six to four per year. Let the cost of these measures be $6 million per year. If that sum is instead spent on something else that does not affect the number of fatalities—for instance, reducing traffic congestion during rush hour—one may say that the value of a statistical human life has been *implicitly* set to be less than $3 million. Again, this is not the value of the life of an identified individual. If someone is trapped in a mine or lost at sea, the cost of the rescue efforts is not a primary issue. The number in our example is also very approximate, since there can be large statistical fluctuations in the number of "saved" lives.

The value of a statistical life can be a useful concept when a society is deciding how to allocate common resources, in particular regarding sea, air, and land transport. Many developed countries assume that it is of the order of US$3 million to $8 million. Accidents that are not fatal may lead to more or less severe injuries. That is also assigned a cost to be used in planning. But we humans are not concerned only with injuries and deaths. On the contrary, we are willing to take risks in order to achieve something we value. For instance, we do not refrain from travelling to visit friends and relatives, in spite of the fact that all travel entails some risk of injuries and death. Therefore, traffic planning may also take into account positive factors, such as a decrease in the time that is spent on commuting between home and the workplace, which can be assigned a certain economic value per saved hour.

There are many ways to assess the value of a statistical life. As we noted above, the decisions by authorities may indicate an implicit value. In the willingness-to-pay approach one can interview people, or one can look at the actual choices made by them. All such studies assume that people are well informed about the consequences of their actions, act rationally, and have a real choice between alternatives. For an individual, the question could be whether to pay for a fire extinguisher at home. Analogous to willingness-to-pay is the concept of willingness-to-accept. We may accept a risky job provided that we get adequate income compensation.

So far we have assumed that the value of a statistical life is always the same. That may not be what most people find reasonable. For instance, let there be a choice between two proactive measures concerning transportation. They are equally expensive and are estimated to reduce an equal number of statistical lives. One of the measures refers to preschool children and the other the elderly. Here, it is likely that priority will be given to the measure that saves the lives of the children. In other words, we have implicitly said that their lives are worth more. With such reasoning, one is considering statistical *years* of life, rather than just statistical lives. But that measure may also be too simple, as the next two examples show. Investing in a pedestrian bridge over a road may save lives. Further, cars can be built so that a serious crash results in severe injuries rather than fatalities. In the first case, lives are saved; in the second there will be injuries instead of fatalities. Both cases add to the statistical years of life, but in one of them it also means living with an injury. Therefore, the concept of quality-adjusted life years (QALY) is often used.

Since there is always more that *can* be done than what we can afford to do, or want to do, assigning a value to statistical lives has been a useful tool in the allocation of resources. Of course, even with this tool, how to allocate resources is still far from a trivial decision. Individuals can have widely differing opinions about, for instance, how one should define quality-adjusted life years. Others rate fatalities differently depending on the circumstances under which the life is lost. Some people may have unrealistic views about how likely certain kinds of fatalities are (for instance, in comparing travel by car and by air) and therefore make ill-informed decisions. Finally, there are those who consider it unethical to give a human life an economic value. In the last case, the consequence can be that scarce resources are not distributed in the best way. Some lives are lost that could have been saved with better prioritization of risk-reducing measures.

Gasoline Here and There

Regular-quality gasoline in the United States and Canada usually has an octane rating of 87, and premium quality has a rating of 91. In Europe the corresponding ratings are 95 for regular and 97 or 98 for the next grade. That raises several questions. Do European cars require gasoline of a higher quality? Can there be fuel with an octane rating above 100?

The octane rating of gasoline measures the resistance of motor fuel to "knocking"—an uncontrolled explosion in the engine. There are two reference fuels—iso-octane (C_8H_{18}) and *n*-heptane (C_7H_{16})—which are assigned octane numbers 100 and 0, respectively. A test sample of the fuel is characterized by octane number, *N*,

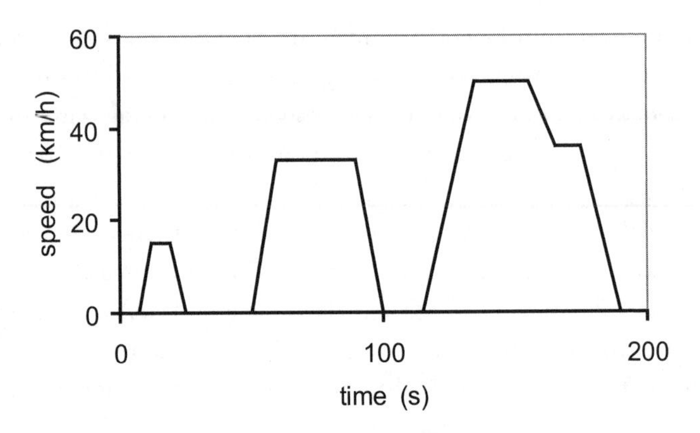

Fig. 2.6. Part of the European urban test cycle to measure fuel consumption of cars

if it has the same knock resistance as a mixture of N % (by volume) iso-octane and $(100-N)$ % n-heptane. The comparison is made in standardized test engines.

The several ways in which such tests are performed lead to somewhat different octane numbers. Most countries, including those in Europe, use the Research Octane Number (RON). Another testing procedure yields the Motor Octane Number (MON). In the United States and Canada, the octane rating is given as the average of the RON and MON numbers. It is four to five points below the RON number. Thus, regular gas in the United States and Canada would correspond to an octane rating of 91 or 92 in Europe. However, octane ratings can vary from region to region, and it is not unusual to find other octane numbers than those just mentioned.

Iso-octane, with RON = 100, is not the most knock-resistant substance. For instance, ethanol has RON = 129 and MON = 102. E85 gasoline (a mixture with up to 85 % denatured fuel ethanol), which is sold as more environmentally friendly, typically has RON = 105. There are additives that can improve knocking properties. One example is tetraethyl lead. Because of its severe environmental effects, it is banned in most countries, and "unleaded" gas is required.

Terms like "premium gasoline" may give the false impression that the gasoline labeled "premium" is always a better fuel than what is termed "regular." But an engine for which regular gas is recommended does not perform better if premium quality is used. On the other hand, if the higher octane rating is recommended, an engine can be damaged if it is run on a lower-octane fuel. At high altitudes, with lower air pressure, engines take in less air. Therefore, the compression of the fuel-air mixture is not as large as at sea level, and an engine can run without problems with an octane number that is one or two units lower than usually recommended.

Soaring gas prices, and concerns about the emission of carbon dioxide and other harmful combustion products, have highlighted measures of how much fuel a vehicle needs. This is not a trivial issue. A simple characteristic quantity is the distance in miles one can drive per gallon of fuel. For instance, 35.5 miles per gallon has been proposed as a federal standard for fuel efficiency in the United States. In Europe, however, fuel consumption is quantified as volume per length instead of length per volume. This is officially expressed in liters per 100 kilometers. Thus, 35.5 miles per gallon is equivalent to 6.6 liters per 100 kilometers.

Records in low fuel consumption are set under idealized conditions, with the vehicle moving at constant speed and on a level and smooth surface. The speed chosen for such tests is an optimum speed that balances the performance properties of the engine and the air resistance. But that is not how vehicles are driven in practice. Outside of built-up areas, speeds may be high and approximately constant, but roads may go over hilly terrain. Driving in cities may be characterized by low speeds and frequent stops. A meaningful measure of fuel consumption should take into account such factors. That requires test runs performed according to a standardized scheme. In Europe such tests are performed on a rolling road, with a test run consisting of an "urban cycle" (average speed 19 km/h, maximum speed 50 km/h, travelled distance 4 km) and an "extra-urban cycle" (average speed 63 km/h, maximum speed 120 km/h, travelled distance 7 km). Japan uses a different but similar test run. Figure 2.6 shows a section of the European urban cycle, with speed versus time.

3

Accuracy and Significance

3.1 Could You Be More Precise, Please?

Why it may be unclear how large the population of Austria is, how much slimmer Maria got in four weeks, and on what day man first took a step on the Moon.

What Is Austria's Population?

A country's census can be a complicated matter, going far beyond the mere counting of the number of people. In Austria it is undertaken only in years ending with 1, such as 2001. Here are four responses to the question, how many people lived in Austria on the census day, May 15, 2001?

- The population was 8 032 926.
- The population was 8 million.
- The population was of the order of 8 million.
- The population was of the order of 10 million.

In some sense all of these statements are correct. The official figure for the number of inhabitants was 8 032 926 — but of course, the population cannot be known to the accuracy of a single person. One trivial objection that may come to mind is the uncertainty in the precise time of birth or death. For instance, some people die in their sleep. A more serious matter is that of defining who is to be included in the count. Many people stay illegally in a country. They may have lived there for years, but still might not be included. Likewise, citizens who live abroad are usually excluded although they may have close ties with their country of origin. The 2001 census in Austria was based on an official registration of where a person's main residence was located. Even that leads to difficulties, and the number of inhabitants was subsequently changed from 8 032 926 to the legally binding result of 8 032 857.

The second statement in our list of answers — a population of 8 million — is correct in the sense that it uses the standardized way to obtain rounded numbers from the result of the census. For instance, numbers from 7 500 000 to 8 500 500 may be given as 8 million. But one cannot be absolutely sure that such a rounding has been done. If someone remembers only that the population was "8 million and something," it is not unusual to say 8 million, even though the actual value might be, for instance, 8.7 million.

In the third statement, the words "of the order of" indicate some uncertainty in the response. The true value may be, for instance, 7 million or 9 million, but definitely not as much as 12 million.

The last statement may express even more uncertainty. Perhaps one knows for sure that it is not as little as 5 million or as much as 20 million. Then "of the order of 10 million" is a reasonable answer. But if the answer comes from someone you know has a good knowledge of the actual number of inhabitants, it is not quite clear how the statement should be interpreted. It could have a meaning such as that in the third answer — that is, 10 million is the correct rounded number, or at least nearly so. It could also mean that the figure will be used in a context where a crudely rounded number gives the relevant accuracy.

Our discussion shows that even a simple question such as that about the number of inhabitants in Austria has several relevant answers. It may be important to judge them in the light of who is making the statement, and for what purpose the information will be used.

A Slim Waist

One should be very suspicious about advertisements promising a slimmer waist in a short time and without effort. What would you think about the following advertisement, found in a European journal (translated here to English)

I lost 3.81 centimeters in just four weeks.
Maria, 38 years

Taking this message at face value implies that Maria could measure the change with exceptional accuracy—to 1/100 of a centimeter (4/1000 of an inch).[1] In this case it is the European copywriter who is to blame for being numerically illiterate. The magic product that claims to make you slim comes from the United States, and the original text says that Maria got slimmer by 1½ inches. That is not very precise. But since 1 inch is exactly 2.54 cm, the formal conversion gives 1.5 inch = 3.81 cm.

Here is another authentic example. In connection with North Korea's nuclear tests in May 2009, a leading European newspaper published a table that gave the range of North Korea's missiles (table 3.1). The seemingly accurate value of 2896 km is conspicuous. This is what one gets from the relation 1 mile = 1609.344 m if the original source says 1800 miles. A better value in the table would have been 2900 km or 3000 km. The journalist just made the conversion without paying attention to the uncertainty that is implicit in the rounded number 1800.

Ridiculously precise numbers may arise not only from the conversion of units but also from simple arithmetic calculations. In an investigation of an accident in the New York City subway, the stopping distance, D, of a train was determined from a measurement of the number of rotations, n, of a test wheel, where $D = nS$ and S is the circumference of the wheel.[2] In the technical report, the circumference was given as 2.911 feet—that is, accurate to almost the width of a human hair. In fact, the diameter of the wheel had been measured as 11⅛ inches (11.125 inches); then that distance had been divided by 12 to get feet and multiplied by π (approximated as 3.14) to get 2.911 041, which was finally rounded as 2.911. If the more accurate number π = 3.141 59 had been used, the circumference would be 2.913 feet. Of course, that is also unreasonable,

Table 3.1. Range of North Korean missiles

Missile	*Range* (km)
Hwasong-5	330
Hwasong-6	700
Rodong-A	1500
Taepodong X	4000
Taepodong 1	2896
Taepodong 2	6000

since the essential uncertainty had already been introduced in the measurement of the wheel's diameter.

In expressions like 3.81 cm and 2.911 feet, it is usually understood that all digits are significant and were often obtained after a rounding procedure. The simple rule is that one should round to the nearest number. For instance, 2.911 is rounded to 2.91 or perhaps to 2.9. Similarly, 2.916 is rounded to 2.92 or 2.9. But what about 2.915, which ends with the digit 5? Should it be rounded to 2.91 or 2.92? If we always choose to take the higher (or lower) value, it may lead to systematic errors in a large set of data. Therefore, the recommendation is to round numbers to get an even last digit. For instance, both 2.915 and 2.925 become 2.92. In some cases—for instance, in dealing with sports records and exposure limits—one must be "on the safe side." The international rules for track and field say that "in all throwing events, distances shall be recorded to the nearest 0.01 m *below* the distance measured if the distance is not a whole centimeter."

The American use of a comma to separate groups of three digits is another potential source of misunderstanding. In documents written in other languages than English, the comma usually means the decimal sign (as we discussed in sec. 1.1, What Is the Point?). Thus, you would write, for instance, $\pi \approx 3{,}142$ (not 3.142). In this case, there is no risk that an American would interpret the value of pi as something beginning with three thousand, but interpretation of written numbers is more problematic when we are dealing with a quantity whose magnitude is not well known. If the text on a food package says "Na 1,260 mg" it is not absolutely clear if this means 1260 mg or 1.260 mg of sodium.

Man on the Moon

In 1969, Neil Armstrong became the first man on the Moon. But when, more exactly, did he step down from the Eagle and utter those famous words: "That's one small step for a man, one giant leap for mankind"?[3] Many people are sure that it was on July 20, but for others on our globe it was already July 21. So how should we give the precise time? Should it be the local time at the space center in Houston, or the local time at the lift-off site in Florida?

There are two main candidates for a generally accepted time—Greenwich Mean Time (GMT) and Coordinated Universal Time (UTC). GMT is simply the local time at longitude 0°, which by definition goes through the Greenwich observatory east of London. It is the local time in the United Kingdom, Ireland, Portugal, and countries in western Africa (except during daylight saving time, where that is used), and it was the basis for the introduction of time zones around the world. GMT was later replaced by the Universal Time (UT), which was also calculated from longitude 0°. In a further refinement, UT was replaced by UTC, which relies on international atomic time. Neil Armstrong touched the surface of the Moon at 02:56 UTC on July 21, 1969, 6½ hours after landing. That is, for instance, 10:56 p.m. EDT (Eastern Daylight Time), July 20, 1969.

In the metric system, the time unit "second" was first defined as 1/86400 of a mean solar day, and then refined to be 1/86400 of the mean solar day on January 1, 1900. The present definition was adopted in 1967. It says that one second is the duration of 9 162 631 770 periods of the radiation corresponding to the transition between two specified hyperfine levels in an atom of the isotope cesium-133 (^{133}Cs).

The length of the so-called tropical year is approximately 365.242

days. It is the time from one equinox to the next. Since this is not a whole number, the day of the equinox, of the summer solstice, and so on would move slowly to other dates in the year if one did not insert leap days every fourth year. However, that is not accurate enough. In 1582, Pope Gregory XIII introduced our present calendar with a new system of leap days. There is one extra day in February each year that is divisible by 4 (e.g., 2008, 2012), with the exception of those years that are divisible by 100 but not also by 400. Thus, the year 2000 was a leap year, but not 1900. The Gregorian calendar was soon adopted in Catholic countries but not until much later in other countries. In the British Empire, including the American colonies, the Gregorian calendar was introduced in 1752, when Wednesday, September 2, was followed by Thursday, September 14.

In addition to the leap year there is the leap second. These are two completely different concepts. The leap year (with one extra day inserted) is a consequence of the fact that the Earth does not rotate a whole number of times around its own axis during its path around the Sun. The leap second, on the other hand, is caused by the slowing down of and small irregularities in the Earth's rotation. The "rotation" (vibration) in atomic clocks is much more stable than the rotation of the Earth. While GMT was defined in relation to the motion of the Earth, its successor, UTC, is now tied to atomic clocks.

In our treatment of Armstrong's first step onto the surface of the Moon, we made no reference to Einstein's theory of relativity, which is so often discussed in connection with space travel. The famous twin paradox says that if one twin travels out from the Earth and then returns, while the other twin remains here, aging has been slowed down for the space traveller. This remarkable consequence of relativity theory is not just a semantic subtlety but a real physical effect. A common objection is that we could equally well consider the traveller as stationary and the other twin to be on the move. Then, as the argument goes, the aging effect should be reversed. This is not correct. The twin who travels undergoes acceleration and retardation, which are felt as forces on the body. Those forces are not acting on the stationary twin. We may compare sitting in an accelerating or breaking car with being in a car that is parked. Therefore, there is no doubt who is travelling and who is stationary. An experiment proving the effect with real twins remains a challenge for future space travel. However, atomic clocks are so accurate that the equivalence of the twin paradox has been proven with clocks carried in commercial aircraft around the Earth.[4]

Finally, the correct time has become an issue in the computer age. With information rapidly spreading on the Internet, it is important to specify when it has been posted. Entries on Wikipedia are posted according to UTC.

3.2 Significant?

On whether you are more likely to flunk if you were born on the 13th day of the month, what constitutes a statistically certain change in voters' opinion, and why you may be uncertain about what time it is if you have two watches.

Flunking

Some students fail in exams. Flunking is so common that it can be worth looking for circumstances that are related to the failures. In such a study one might obtain the surprising result that students born on the 13th day of the month run a statistically proven increased risk of failure in their studies. Does this seem reasonable? Of course not, but nevertheless it may be what the mathematical analysis showed. Perhaps the samples of students were too small, so that statistical fluctuations affected the result? That could be, but if the survey is repeated with a very large number of students, one might instead find that those born on the 19th day of the month are more likely to fail — again a statistically certain result.

The problem lies in the concept of "statistically certain." Assume that we suspect a chemical X to be harmful and cause cancer. However, not all those exposed to X develop cancer. Further, cancer is also common among people who have not been in contact with X. We now take a large statistical sample of the population and divide it into two groups. One group contains only those who have been exposed to X. Among them N_C people are found to contract cancer. Based on the frequency of cancer in the other group, we expect that number to be N_0 if the suspected chemical is harmless. If N_C is larger than N_0 it may indicate that X is carcinogenic. Suppose that the result is $N_C = 1720$ and $N_0 = 1709$. Such a small difference is easily caused by statistical variations in N_C and N_0. On the other hand, a result that $N_C = 2311$ and $N_0 = 1709$ gives strong evidence that X is carcinogenic. How large must the difference between N_C and N_0 be for such a conclusion?

Statisticians would usually say that an effect is statistically "proven" if there is less than a 5 % chance that the result is due to statistical fluctuations alone. It follows that typically 1 in 20 investigations (i.e., 5 %) will find a connection between an effect and its possible cause, even if there is no such relation whatsoever. This problem is not solved with very large samples, because the criterion of 5 % due to chance alone is independent of the sample size.

In our introductory case of exam failure there are 31 numbers for the days in a month. Therefore, there is a good chance that at least one correlation between exam failure and a particular date of birth will be found to be "statically proven" at the level of 5 %.

Investigations looking for correlations between diseases and agents that may cause the diseases are called epidemiological studies. If one uses the "only 5 % due to chance" criterion, it lies in the very nature of the methodology that there will be some false alarms. This is referred to as the *mass significance problem*. When the study is repeated, by the same researchers or other groups, the false alarms will be identified as such—provided, of course, that there is not something dubious in the way the study was performed. So-called *confounding factors* may be very difficult to eliminate.

A confounding factor affects the outcome of an investigation and can lead to erroneous conclusions. For instance, suppose that an epidemiological study found a positive correlation between drinking coffee and getting lung cancer. There is a well-established correlation between smoking and lung cancer, so it could be that those who drink coffee are more likely to smoke than those who don't drink coffee.

Of course, there can be incorrect relationships without the need to identify a confounding factor. For instance, it has been noted that there is a correlation between the number of breeding storks and the birth rate in European countries,[5] thus "confirming" the folklore that babies are delivered by storks!

A Change in Opinion

In modern society, voters' preferences are monitored in polls conducted by various companies. The survey methods these organizations use may differ in details, but basically the question is the same: What would be the outcome of an election or a referendum if it were held today? Mass media report the results, often with the comment that a particular change is statistically significant, or that it is within the margin of error or statistical uncertainty. What is meant by that?

As an example, take a case in which there are three political parties or three alternatives, A, B, and C. To make the point and have simple calculations, we will assume that there are exactly 1000 responses in the poll, with the number of supporters and their share (percent) of the votes as shown in table 3.2. The next time the poll is conducted, the results are those shown in the right-hand side of the table.

Interpreting these results is far from easy. It might seem at first as if 2.5 % of the voters changed from A to C. However, that is the net outcome of the poll. In reality, it could be that, for instance, 2.5 % of the voters changed from A to B and the same number changed from B to C. This is just one of a large number of possibilities. While the detailed flow of voters between the alternatives is important for the parties' strategies, the public is more interested in the final result. Are the changes statistically significant in comparison with the previous poll? Alternative A lost 25 votes among an expected number of 500 (the number in the first poll); thus, it seems that only 1 in 20 supporters changed his or her mind. This is in contrast to alternative

Table 3.2. The results of two hypothetical polls

	First poll		*Second poll*		
Alternative	*No. of votes*	*% of votes*	*No. of votes*	*% of votes*	*% change*
A	500	50.0	475	47.5	−2.5
B	400	40.0	400	40	0
C	100	10.0	125	12.5	+2.5

C, where there were 5 new supporters for each group of 20 in the first poll. We may get a feeling that the increase in support for C can be characterized as statistically significant, while the loss for A perhaps lies within the margin of error.

How one should define "statistically significant" or "within the margin of error" is a very complicated matter. In our example, the sample for the two polls would normally consist of different people. Perhaps the result for A in the first poll was high because that sample had more supporters of A than usual, due to a statistical fluctuation. It could also be that the sample in the second poll had fewer supporters of A than usual, again because of a statistical fluctuation. That could explain the whole apparent loss for alternative A. Therefore, one must be very cautious in the interpretation of changes between two consecutive polls. Trends revealed in a long sequence of polls are more reliable. Those performing the poll may also use more sophisticated techniques than just counting the number of votes for the alternatives.

One can work out theories for the expected statistical variations in polls and find how likely it is that a change is due to chance alone. It is usually considered as acceptable that the "true" result for a certain alternative may lie outside the margin of error in 5 % of the cases. Table 3.3 shows how these margins vary with the percentage support, according to a standard mathematical approach,[6] when the poll contains 1000 responses. The largest margin of error, ±3.1 %, is obtained for an alternative that attracts half of all voters. In table 3.2 the drop in the support for A was 2.5 %, which therefore lies

Table 3.3. Margins of error in a poll with 1000 responses

Total support, %	10	25	50	75	90
Margin of error, %	±1.9	±2.7	±3.1	±2.7	±1.9

within the margin of error. Alternative C gained 12.5 % of the voters. In that case the margin of error is only about ±2.0 % (interpolated in table 3.3 to 12.5 % share of all votes), so the result is a "statistically certain" increase for C.

It must be stressed again that the margin of error is not a unique concept but depends on how much we choose to restrict the possible role of chance alone. That is a matter of convention and can vary with the type of study. Of course, the uncertainties get smaller if more people are included in the poll. In order to reduce a certain margin of error to half its value, the polling organization must make the sample four times larger. With 2000 people in the poll, the margin of error for alternative A drops from ±3.1 % to 2.2 %.

Error Bars

If you are asked what the outdoor temperature is, you may look at the thermometer and say 59 °F, or 15 °C. It is then understood that this is what you read from the scale. Another person reading from the same scale may find that it is closer to 60 °F and report that value. It seems that the temperature can be determined with an uncertainty that is at most one mark on the scale—for example, ±1 °F. But your neighbor may tell you that according to her thermometer, it is two degrees warmer than what you said. Of course, it is possible the actual temperature is different, but it could also be that the two thermometers show different values even if their surrounding has the same temperature. Remember the old saying that a man with one watch knows what time it is, but a man with two watches is never quite sure!

Scientists distinguish between the *uncertainty* and the *accuracy* of

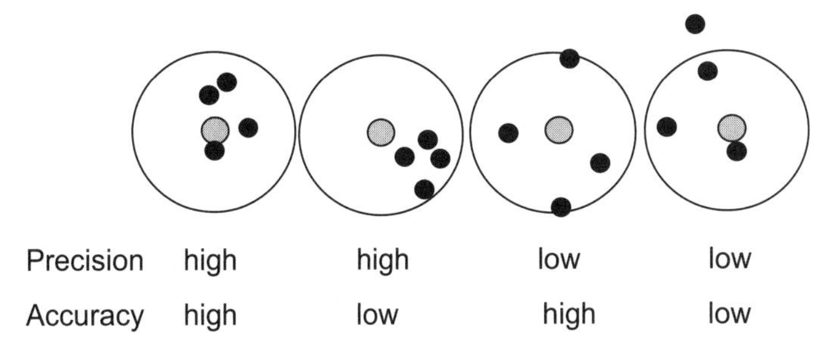

Fig. 3.1. Illustration of the concepts of precision and accuracy

a measurement. The uncertainty in a measured value is often expressed in numbers, for instance as in the expression (60 ± 1) °F. Accuracy is a qualitative concept and is not given a numerical value. An accurate measurement gives a value that is close to the true value. Further, one can distinguish between accuracy and *precision*. Figure 3.1 illustrates this difference.

The basic physical theories rely on only a few constants of nature. Arguably the three most fundamental of them are the speed of light in vacuum (c), which appears in relativity theory; Planck's constant (h), which governs phenomena in quantum mechanics; and the Newtonian constant of gravitation (G). Physicists have tried to determine these constants as accurately as possible. After a careful assessment of experimental data, it was concluded in 1957 that the speed of light in vacuum is $c = 299\,792.5 \pm 0.4$ km/s. Not much later, in 1983, scientists decided to give it an *exact* value by defining the meter as the distance that light travels in vacuum in 1/299792458 of a second. The value of the Planck constant is $h = 6.626\,068\,96\,(33) \times 10^{-34}$ Js (joule second), according to an assessment published in 2006. Here "(33)" is the standardized way to express uncertainty, which in this case would often be written $\pm 0.000\,000\,33 \times 10^{-34}$ Js.

In sharp contrast to the exact value of c and the accurate value of h, the value of the gravitational constant, G, is not well known. It had already appeared in Newton's theory of gravitation, which says that the force between two objects with masses M_1 and M_2 and separated by the distance R is

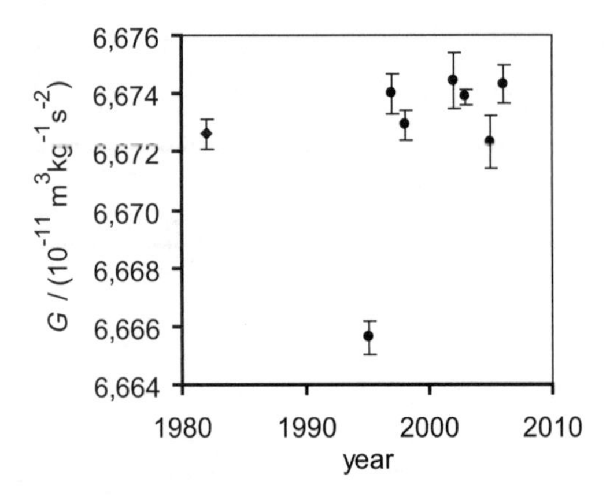

Fig. 3.2. Some recent determinations (among many) of the Newtonian constant of gravitation. The last entry, for 2006, is the CODATA recommended value.

$$F = \frac{GM_1M_2}{R^2} \quad .$$

Sometimes G is called "big G" to distinguish it from the closely related "little g," which is the acceleration of gravity on the surface of the Earth and is known from its approximate schoolbook value, 9.8 m/s^2 or 32 ft/s^2. Big G is notoriously difficult to measure. Its first measurement was provided by Henry Cavendish more than 200 years ago. Since then, there have been a large number of attempts to determine the Newtonian constant of gravitation. The recommended value from CODATA (Committee on Data for Science and Technology) 2006 is $G = 6.674\,28\,(67) \times 10^{-11}\ \mathrm{m^3\,kg^{-1}\,s^{-2}}$.

When scientific data are presented in graphs, their uncertainties are usually shown as error bars. However, assigning an error bar can be a rather subjective matter. If several measurements are taken for the same quantity, the statistical spread gives an idea about the *random* errors, that is, about the precision of the measurement. It is much more difficult to estimate the *systematic* errors, and often the error bars are nothing but the personal judgment of the scientist.

Figure 3.2 shows some recent measurements of G, and also the

CODATA value recommended by NIST (National Institute of Standards and Technology) from 2006.[7] We see in figure 3.2 that not all error bars are mutually consistent. An error bar is *not* an absolute limit outside of which a value cannot lie but expresses the likely uncertainty.

3.3 Limit Values

When loud music is too loud, whether one milligram of polonium can kill a man, and whether you should worry about snapping elevator cables.

Will Your iPod Make You Deaf?

Can listening to music make you deaf? Legislators around the world are taking this problem seriously. It is obvious that we should not be subject to sound, or noise, that is too loud. But how loud is too loud? Sound levels are measured in decibels, often written dB(A). The label A means that sound pressure is weighted according to the sensitivity of the human ear to different frequencies. Like so many other measures of dangerous levels, there are no objective sharp limits. It is generally agreed that both the absolute sound level and its duration must be considered. This is reflected in recommendations and legislation. For instance, the US federal agency NIOSH (National Institute for Occupational Safety and Health) has arrived at the following maximum exposure times:

Sound level	*Maximum exposure time*
85 dB(A)	Less than 8 hours
110 dB(A)	Less than 1½ minutes
140 dB(A)	Can damage hearing after a single exposure

In the European Union member states have been required to comply with European Directive 2003/10/EC, for control of noise at work, since 2006. A lower and a higher level are defined as averages over time. At or above the lower level, 80 dB(A), employers are

Table 3.4. Typical noise levels

Activity	*Noise level,* dB(A)
Breathing	10
Whispering, 1.5 m away	20
Rainfall	50
Normal conversation	55–60
Busy street	80–85
Heavy truck, 7 m away	95–100
Pigpen at feeding time	100–110
Shouting in the ear	110
Personal music player on high	110
Rock concert	110–120
Jet taking off 25 m away	140

obliged to make suitable hearing protection devices available on request, but they cannot enforce their use. At and above the upper level, 85 dB(A), employers must strictly enforce the use of hearing protection. Moreover, there is an exposure limit value of 87 dB(A) at the ear while the employee is using the hearing protection equipment. These rules apply to all employees who are subjected to sound or noise, including musicians and people employed in the entertainment sector. It is interesting to compare the numbers that NIOSH and the EU legislators decided upon with typical noise levels in various activities, shown in table 3.4.

The decibel scale is logarithmic. An increase by 3 dB is perceived as a doubling of the noise. If two sources simultaneously emit sound at the same level, their combined effect corresponds to an increase by 3 dB. Further, it is usually assumed that a doubling of the exposure time has the same effect as an increase in the sound level by 3 dB. We can check that the NIOSH limits given above—85 dB(A) for less than 8 hours and 110 dB(A) for less than 1½ minutes—are consistent with these principles. Each increase in the noise by 3 dB(A) requires halved exposure time. Taking eight steps of 3 dB(A), from 85 to 109 dB(A), means that the exposure time of 8 hours (= 480 minutes) must be reduced by a factor of $2^8 = 256$, giving somewhat less than 2 minutes. The remaining increase by 1 dB(A) brings the exposure time down to about 1½ minutes.

Returning to the hearing risk associated with iPods, it depends on many factors, among them age and the kind of headphones used. There is general consensus, however, that one should not listen at the highest iPod level for more than about 5 minutes per day.

Lethal Dose

Which snake is the most venomous, and which chemical is the most poisonous? Those questions are not easy to answer. It is natural to assume that even if a large dose of something is lethal, an extremely small dose will not cause any noticeable harm (although we will see an exception to this with the linear no-threshold hypothesis with regard to radiation in sec. 4.2). Further, it is likely that not all individuals are equally sensitive. A small person may react more severely to a substance, since the concentration in the body will then be higher. When a large population is subject to a lethal agent, we can plot the frequency of fatalities in percent versus the dose expressed as the amount, X, of the agent divided by the mass, M, of the individual. Figure 3.3 shows a typical curve. The value of X/M where the lethality is 50 % a certain time after the exposure is usually taken as a measure of the lethal dose, denoted LD50 or LD_{50}.

Of course, LD50 values are not well known for humans, since we can't perform controlled experiments. The published values are often inferred from experiments on animals like rats, but they may react very differently from humans. For instance, the Australasian funnel web spider is one of the most venomous spiders and caused many deaths in Australia before an antivenom was developed in the 1980s. Yet, a bite from the funnel web spider does not kill horses, rabbits, dogs, or cats. Another important factor is the way a toxic agent is brought in contact with the body—for instance, whether it

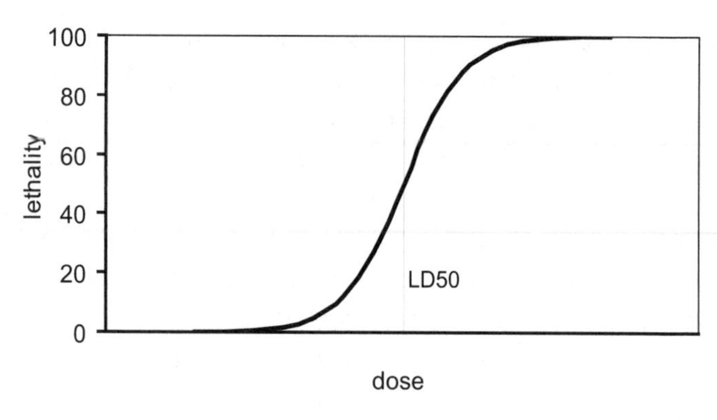

Fig. 3.3. Lethality, in percent, as a function of dose

is inhaled, swallowed, or taken up directly by blood vessels. It is obvious that lethality versus X/M curves such as that in figure 3.3, and lists of LD50 values such as those provided in table 3.5, have large uncertainties.

The Bulgarian dissident Georgi Markov became the victim of one of the most remarkable murders by poison. On September 7, 1978, Markov was waiting at a bus stop on Waterloo Bridge in London when he was suddenly jabbed in his calf by a man with an umbrella. Markov died a few days later. In a postmortem examination it was discovered that he had been shot with a pellet gun. A small sphere made of platinum and iridium, only 1.5 mm in diameter, had bore holes 0.28 mm in diameter containing highly poisonous ricin. The case was never completely solved.

The concept of lethal dose refers not only to chemicals but also to radioactivity. A famous example is the death of the former Russian Federal Security Service officer Alexander Litvinenko, who became the victim of the radioactive element polonium-210 (^{210}Po). Litvinenko had received asylum in Great Britain. On November 1, 2006, he suddenly fell ill, and on November 23 he died at the University College Hospital in London. It was first thought that Litvinenko was poisoned with thallium, but soon the lethal substance was identified as polonium.

Unlike most other radioactive substances, ^{210}Po emits only alpha

Table 3.5. Lethal doses, LD50, for selected venoms and poisons

Venom or poison	*LD50* (mg/kg)
Western diamondback rattlesnake	2
Cobra	0.5
Curare	0.1
Fierce snake (inland taipan)	0.025
Ricin	0.003
Polonium-210	0.00001

particles. They cannot even penetrate a sheet of paper, but once inside the body, polonium is extremely harmful. From the known effect of a radiation dose, it has been estimated that 1 microgram of ^{210}Po is sufficient to cause death. The investigation by British authorities revealed a remarkable pattern. Even extremely small amounts of ^{210}Po leave a trail that can be accurately followed. Two former KGB and GRU officers who had met with Litvinenko at restaurants in London at least twice could be linked to the substance. Traces of polonium ^{210}Po were found in a teacup in one of the restaurants. Further, one of the officers had been on British Airways flights from Moscow to Heathrow and back. British Airways later published a list of 221 flights that had used the contaminated aircraft. The request by the British Foreign Office to the Russian Government to extradite one of the suspects was declined.

The Weakest Link

When you buy a rope, there may be a tag attached which says, for instance, "450 kg" or "1000 lb." Ignoring the fact that this is a mass, and not a force, the meaning is obvious: The rope will not break under a load smaller than 450 kg. Similarly, a chain might be labeled "10 000 kg." However, there is an essential difference between the rope and the chain, apart from the difference in strength.

No material is free from defects, and these defects determine its strength. It makes no difference if we think of a piece of soft iron or of superhard steel. They will have different strengths, but in both cases failure depends on those inevitable imperfections. The very concept of defects implies that there are variations even among pieces that seem to be identical. Therefore, not all links in the chain will yield under exactly the same load — one of them is weaker than the rest. If the load on the chain is gradually increased, the failure comes suddenly when the weakest link snaps, while the rest of the chain appears intact.

A rope is usually made of a large number of parallel fibers that are intertwined in a characteristic spiral pattern. The stress that an individual fiber can take varies, just as was the case with the metal links. Further, the fibers are held together in such a way that they will not all carry exactly the same share of the total load. When the load on the rope increases, there will be one strand that fails first under the stress. The initial failure could be a break in the fiber itself or in its connection to adjacent parts of the rope. Whatever the failure mode is, the remaining fibers must carry a slightly higher load. Next, another fiber snaps and leaves a larger burden to the rest.

Ultimately the rope breaks, but unlike the case with the chain the break does not occur abruptly. Instead, there may be signs like visible broken strands or a larger-than-expected elongation. The rope has a kind of what engineers call *redundancy* — one particular failure will not immediately jeopardize the function of the entire system. Redundancy is a key concept in engineering practice, in particular when safety aspects are of concern. In a car with dual independent braking systems, the braking performance is not perfect if one system fails, but the malfunction is likely to be detected, and a catastrophe is avoided since substantial braking power is still available.

In an elevator there is usually a sign saying, for instance, "Capacity 8 persons." This is a kind of limit value, but of course a load of eight people is not anywhere near the breaking strength of the elevator cable. The total weight of eight people can vary a lot, and they may carry luggage. Further, the load of the cable is increased when the elevator accelerates, according to Newton's law. As we have just

Table 3.6. Safety factors for elevators in California

Rope speed		*Safety factor*	
ft/min	m/s	*Passenger*	*Freight*
200	1.0	8.40	7.45
600	3.0	10.70	9.50
1000	5.0	11.55	10.30
1500	7.6	11.90	10.55

noted, the breaking strengths of elevator cables are not the same, even if they are manufactured to be identical in their properties. The engineer who designed the elevator must also allow for accidental variations in installation and maintenance and for various human errors.

As a consequence of all these uncertainties, there are building codes that prescribe what are often termed *safety factors*. The design must accommodate a load that is larger than the nominal allowed load by a certain factor, in the case of elevators often about 10. Table 3.6 gives examples of such safety factors for California. If worse comes to worst and the elevator cable snaps—for instance, because of sabotage—there is still at least one redundant system that will stop the fall of the elevator. The name to think of is Elisha Graves Otis. In 1854, at New York's Crystal Palace exhibition, he performed a famous experiment in front of an enthusiastic crowd (fig. 3.4). As the cable was cut, the fall of the elevator was almost instantaneously halted by a safety mechanism.

Fig. 3.4. The celebrated demonstration by Elisha Graves Otis of his elevator safety mechanism

3.4 Fair Games?

Whether the lanes in an Olympic swimming pool are equally long, why you might have a shorter time in 100-m dash with manual timing, and whether it is easier to set a triple jump record in Brazil than in Germany.

Winning by a Small Margin

The 5000-m world record in racing for men has been broken several times with a very small margin. The improvement was less than 1 second on four occasions before 1966. Table 3.7 shows more recent world records. In 1972 the new record was better than the previous record by 0.2 s, in 1977 by 0.1 s, and in 1985 by a tiny 0.01 s. The distances covered by the athletes during these short time intervals were about 1.2 m, 0.6 m, and 0.06 m, respectively. Thus, it is interesting to compare how much various sports venues may differ from a distance of exactly 5000 m.

The rules of the International Association of Athletics Federations (IAAF) from 2008 give a very detailed description of how a standard track is to be measured. Its length must lie in the interval 400.000 m to 400.040 m, at a distance of 30 cm from the inner curb. Thus two tracks at different venues can differ in length by at most 4 cm. The measurements are now done with a theodolite. Earlier rules said that one could use steel measuring tapes. Two separate measurements were then required, whose results must not differ by more than $(0.0003L + 0.01)$ m, where L was the measured distance in meters. If we take this expression as characteristic of the maximum error for one lap (400 m) in the stadium, we get 13 cm. The 5000-m event is run in 12.5 laps. Even with the presently allowed difference of 4 cm per lap, we get an accumulated distance of 12.5×4 cm $= 0.5$ m.

At the 1972 Olympic Games in Munich, the swimmer Gunnar Larsson, Sweden, won the 400-m medley with the time 4.41.981, beating Tim McKee, USA, who got 4.41.983. The difference in the official times was only two-thousands of a second. During that short

Table 3.7. Some world records in 5000-m race for men

Athlete	*Record* (min)
Ron Clarke, Australia (1966)	13.16.6
Lasse Virén, Finland (1972)	13.16.4
Emiel Puttemans, Belgium (1972)	13.13.0
Dick Quax, New Zealand (1977)	13.12.9
Henry Rono, Kenya (1978)	13.08.4
Henry Rono, Kenya (1981)	13.06.20
David Moorcroft, Great Britain (1982)	13.00.41
Said Aouita, Morocco (1985)	13.00.40
Said Aouita, Morocco (1987)	12.58.39
(6 new records 1994–1997)	
Daniel Komen, Kenya (1997)	12.39.74
Haile Gebrselassie, Ethiopia (1998)	12.39.36

time one would swim almost 3 mm. In 400-m medley, the swimmers cover the 50-m lane eight times. The margin by which Gunnar Larsson won thus corresponds to a difference by less than 0.4 mm in the length of a lane. It is interesting to compare the setup for the Olympics with the FINA (Fédération International de Natation) rules. They require the length of a lane to be between 50.00 m and 50.03 m, at all points from 0.3 m above to 0.8 m below the surface of the water. These tolerances may not be exceeded when touch panels are installed for automatic timing. In the 1500-m event, contestants swim the lane 30 times. Thus, a difference of 3 cm in the length of a lane corresponds to a total time of about 0.5 s.

It may seem unreasonable to measure the times as accurately as to within 0.001 s. The rules were also changed and now read: "Timing shall be to 2 decimal places (1/100 of a second)." With this new rule, the gold medal would have been shared between Larsson and McKee. In the 100-m freestyle swimming event at the 1984 Olympic Games in Los Angeles, Nancy Hogshead and Carrie Steinseifer, USA, both got the time 55.92 s and shared first place.

Accurate Timing

Few world records in sport were broken during World War II, for obvious reasons. But Sweden was neutral, and several new records were set there. One famous example is Gunder Hägg's 1500-m record time of 3.43.0. In those days hand timing with stopwatches was used. The accuracy was at the very best 0.1 s. The rules said that for a record to be acknowledged, there had to be three official timekeepers. If two of three watches agreed, and the third disagreed, the time recorded by the two was the official time. If all three watches disagreed, the middle result was official. When Gunder Hägg set his record 1944 in Gothenburg, the three watches showed 3.42.0, 3.43.0, and 3.43.1, respectively.

In the 1960s, electronic timing became increasingly common, and since 1977 it has been the only method accepted by IAAF for world records. The timing starts automatically with the starter's gun. The finish is recorded through a camera with vertical slit. The image is synchronized with a timescale graduated in 1/100s of a second. Times are read from the recorded photo-finish image for the moment when the torso reaches the vertical plane at the finish line.

Figure 3.5 shows the 100-m world record for men. The last record with hand timing is from 1960. The straight line, which is meant only as a guide to the eye, suggests that the automatically recorded times (circular dots) are longer. This may at first seem surprising but has a simple explanation. The athletes start at the sound of a gun, fired close to them. The velocity of sound in air is about 330 m/s, so it takes about 0.3 s for the sound to travel 100 m to the finish line. For that reason, the stopwatches could not be started when the sound was heard, but at the sight of smoke or flame coming out of the starter's gun. Because of human reaction time, the watches were nevertheless started somewhat late. But at the finish line, the timekeepers could see the approaching athletes and stop the watches without the reaction time delay.

This effect is recognized in the 2009 IAAF scoring tables for athletics, which mandate that when hand timing is used, officials should

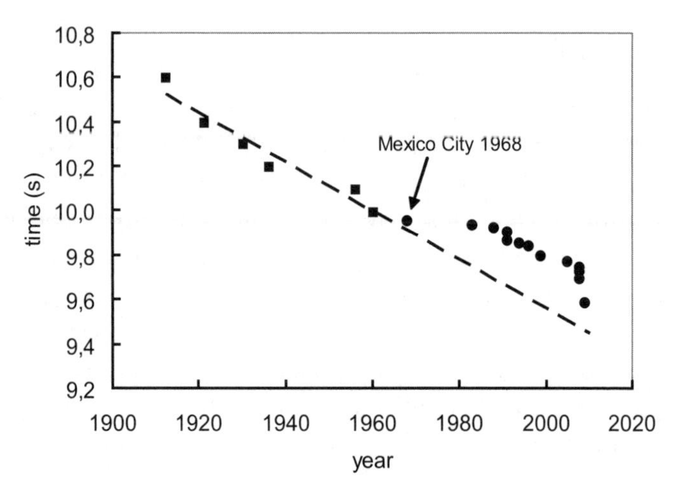

Fig. 3.5. Records for men's 100-m race. Squares are results obtained with hand timing and circles with automatic timing.

add 0.24 s in sprints and hurdles up to 200 m, and add 0.14 s for 300-m, 400-m, and 400-m hurdles. The first automatic timing point in figure 3.5 may seem inconsistent with our discussion. However, that is the result from the 1968 Olympic Games held in Mexico City, where high altitude, leading to low air drag (air resistance), was one reason for improved results in sprint events.

Going back to the 400-m medley of the 1972 Olympics, it may not be correct to say that the difference between Larsson and McKee corresponds to a margin of 3 mm. All race results must be given as rounded numbers, in one way or another. This is a problem one cannot avoid, even if the measuring equipment has infinitely high accuracy. Suppose that two competitors are automatically registered with the times 50.0001 s and 50.0050 s. The difference is 0.0049 s. Further, suppose that there is a rule saying that a race is regarded as a tie if the difference is smaller than 0.0050 s. In this case they will share the position in the final result. But if the registered time 50.0050 is increased by only 0.0001 s, it is no longer a tie. Electronic timing, where competitors break a laser ray or touch a panel, allows decisions about whether there is a tie to be completely automated.

Even an infinitesimal difference can tip the balance in either direction. The problem remains also when the time is recorded not by clocks but on film, as with photo-finish equipment. Naively, one might say that in case of doubt, one should judge it to be this or that result. Alas, there is no sharp distinction between being in doubt or not.

In downhill skiing official times are given to a precision of 0.01 s. The speed at the finish typically is 20 m/s. If the winning margin is 0.01 s, a newspaper may rephrase this as indicating that the margin was only 20 cm, or less than one foot. But the official time difference of 0.01 s can be anything between 0.00 s and 0.02 s, depending on rounding effects. For instance, if the rounding rule says that one goes to the nearest higher hundredth of a second, the actual times 85.0001 and 85.0099 would both be reported as 85.01 s. In contrast, the times 84.9999 and 85.0101, which also differ by very nearly 0.01 s, would be rounded to the clearly separated results 85.00 s and 85.02 s, respectively.

Are All Sports Venues Equivalent?

At the 1936 Olympic Games in Berlin, the Japanese athlete N. Tajima won the triple jump with 16.00 m. This is one of the classic track and field records. In 1950, A. Ferreira da Silva from Brazil also jumped 16.00 m. The next year he improved the record to 16.01 m. But Ferreira da Silva's results are from São Paulo and Rio de Janeiro, respectively. In those two cities the acceleration of gravity is 0.25 % lower than in Berlin (table 3.8). This fact gave rise to a discussion of its effect on the triple jump record. In analogy to the result of a simple model for the long jump,[8] one would expect that Tajima's result in Berlin would have been a few centimeters longer in Brazil.

The main reason for the variations in the acceleration of gravity shown in table 3.8 is that the Earth is nonspherical, with a radius that increases by 0.34 % (21 km) from the poles to the equator. The additional decrease of g due to the altitude in Mexico City (2.3 km) is almost negligible. A much more important aspect is the low air

Table 3.8. Acceleration of gravity at various sports venues

Venue	*Latitude* (°)	g (m/s^2)
Berlin	59	9.813
Rome	42	9.803
Los Angeles	34	9.796
Mexico City	19	9.779
Rio de Janeiro	23	9.788
São Paulo	24	9.789
Melbourne	38	9.800

pressure, and therefore decreased air drag, at high altitudes. This is particularly relevant for the famous long jump record, 8.90 m, set by Robert Beamon at the 1968 Olympic Games in Mexico City. The improvement over the previous world record was a staggering 55 cm.

We can get an idea of the effect of high altitudes if we think of the rule about the maximum allowed wind speed. It must be less than 2 m/s in the direction of the jump. The air resistance is proportional to the air pressure (i.e., the air density) and increases as the square of the speed relative to the air. During Beamon's jump the wind speed was the highest allowed, 2 m/s. If we assume that the speed of the athlete is 11 m/s, then the air resistance is reduced by a factor of $(11 - 2)^2/11^2 = 0.67$, or by 33 %. The air pressure in Mexico City is about 23 % lower than at sea level. Thus, a maximum allowed tail-wind is somewhat more important than the effect of high altitude and no wind. If we simultaneously have an air pressure reduced by 23 % and the maximum allowed wind speed of 2 m/s, the combined effect is a reduction of the air resistance by about 40 % compared with still air at sea level. A low air resistance means that the speed at takeoff is higher and the in-flight drag force is lower. Simple models indicate that the combination of wind and high altitude accounts for a large part of the record jump in Mexico City. While lower air pressure will tend to improve results in long jump, triple jump, and sprint events, the effect is the opposite in endurance sports, which rely on the body's oxygen uptake.

We saw that the variation in gravity may have a small but not entirely negligible effect in triple jump. Its effect is more complicated in shot put and hammer throw. With lower gravity the shot feels a bit lighter when you lift it, and for a given release speed it will go farther. But it is the weight of the shot, and not its mass, that has changed. The inertial resistance that the athlete must overcome to give the shot its release speed depends on the mass and not on the force of gravity. That is independent of the venue.

4
Extrapolations

4.1 The Dangerous Exponential

How to look like a mathematical genius, get more wheat than there is on the Earth, and tap a finite resource forever.

The Rule of 72

At a meeting you read in the handouts that a certain activity is projected to increase steadily by 6 % per year. Some people in the meeting are having difficulty grasping what that means in practical terms, and so you say in the following debate, "Twelve years from now, it will have grown to double the size of today." You don't need to be a mathematical genius to reach that conclusion. Very simple mental arithmetic suffices. For instance, if the annual increase had been 4 % instead, it would take 18 years for a doubling in size. This is a simple application of *the rule of 72*. It says that if there is a steady annual increase by p %, the doubling time (the number of years needed to increase by a factor of 2) is

$$\text{doubling time} \approx \frac{72}{p} .$$

If something *decreases* by p % annually, the time needed for it to be halved is $72/p$ years. The rule can also be reversed so that we can ask, What annual change, p, in per cent is needed to get something doubled, or halved, after y years? The answer is

$$p = \frac{72}{y} .$$

The rule is not mathematically exact, but it is a very good approximation, as shown in table 4.1.

The number 72 has many simple factors:

$$72 = 2 \times 2 \times 2 \times 3 \times 3.$$

Therefore, we can easily divide 72 by any of the numbers 1, 2, 3, 4, 6, 7.2, 8, 9, 10, 12, and 18 and get the doubling times corresponding to such annual increases in percent. Sometimes the method is called the rule of 69, and the doubling time is calculated as $69/p$. This is mathematically more correct when p is very small, but 72 is much more practical to use, and anyway the rule is only approximate.

As an application, consider the world's population. In the beginning of the twenty-first century it has been increasing by 1.2 % annually, so it will be doubled in $72/1.2 = 60$ years if the growth rate remains the same. Some countries have a decreasing population. Suppose that in these countries, the population shrinks by as much as 1 % annually. Then it will be halved in about 70 years.

We can take another example from today's environmental issues. A politician says that the use of a certain pollutant must be halved in 20 years. You can quickly calculate that achieving this goal requires an annual decrease by about $72/20\ \% = 3.6\ \%$. After another 20 years, at the same rate of decrease, the level will be down to one quarter of today's value.

Finally, note that doubling 10 times means an increase by about a factor of 1000, since $2^{10} = 1024$. This is a simple and useful result, with a factor that we recognize from the concept of one kilobyte (table 1.3).

A Problematic Reward

The legend about the origin of the game of chess has many different versions. In one of them, an Indian named Sissa presented his invention to the king. The king was so impressed that he asked Sissa what he would like to get as a reward. Sissa was a very clever man and

Table 4.1. Doubling times for a given annual increase, p (%)

	2%	4%	6%	8%	10%	12%	18%
Doubling time (years), with rule of 72	36	18	12	9	7.2	6	4
Doubling time (years), more accurately	35.0	17.7	11.9	9.0	7.3	6.1	4.2

said, "Give me one grain of wheat for the first square of the chessboard, two grains for the second square, four grains for the next square, and so on. Each time the number of grains should be doubled." The king gladly accepted this proposal and asked his servants to bring a sack of wheat grains to fulfill Sissa's wish. But by the time they came to the 20th square, the sack was empty, and the next sack was just sufficient for the 21st square. The king realized that his land did not have enough wheat to give Sissa his reward, but the king was also a clever man. He told Sissa to count every grain of wheat, in order to check that the right amount was given.

It is instructive to consider the first steps in the sequence, and also to sum up the total number of grains after each step. Table 4.2 shows the result for the first 10 squares. Notice that the accumulated number of grains after each step is one grain less than the number of grains to be placed on the next square. It soon becomes an extremely good approximation to say that the two numbers are equal, and we can formulate the following very useful rule:[1]

In a long and steady exponential growth, the accumulated amount up to a certain point is equal to the amount added during the next doubling period.

Returning to the chessboard problem, the number of grains on the 64th square would be 2^{63}. The total number of grains on the board is twice as much, or $2 \times 2^{63} - 1$, to be exact. This is approximately 10^{19}, or roughly equal to 1000 times the world's present annual production of wheat.

Table 4.2. Number of grains on the first 10 squares on the chessboard

	Chessboard square									
	1	2	3	4	5	6	7	8	9	10
Number of grains	1	2	4	8	16	32	64	128	256	512
Sum of grains	1	3	7	15	31	63	127	255	511	1023

The impossibility of continued exponential growth is obvious but still difficult for many to fully comprehend. An annoying example is the sending out of chain letters. A typical chain letter can look like this:

> *Send a postcard to each of the five persons whose names and addresses are on the top of the list in this letter. Then copy the text of the letter, but delete the list's top name and add your own name and address at the bottom. Send copies of the letter to five people who are not on the list. You will soon receive thousands of postcards.*

The idea is that your own name starts at the bottom of the list in the five letters you send out. The next time someone sends out postcards your name has moved to second from the bottom and it appears in $5 \times 5 = 25$ distributed letters. When your name has reached the top, it has appeared in $5 \times 5 \times 5 \times 5 \times 5 = 3125$ letters. If no one breaks the chain, these 3125 persons will send you a postcard. It seems to be a brilliant idea. Each person participating in the chain sends only five letters and five postcards but receives 3125 postcards in return.

Surprisingly many people are fooled by this trick. Or at least they think that if they are among the first to enter the chain, there is a good chance that it will continue with many links before it collapses. When that happens they are out of the game, but not before they have received all the postcards. Even more surprising is that people fall for the same trick when money is involved, perhaps more cleverly disguised in pyramid schemes. Pyramid schemes are illegal in many countries, but the case of Bernard Madoff shows that people may never learn.

Suddenly Nothing Was Left

A well-known problem in popular mathematics can go like this:

> *In my mother's garden there was a pond with water lilies. Each year they increased to twice as many as before. In 1987 the pond was completely covered by the water lilies. When was the pond half covered?*

The expected answer is 1986. This problem may seem trivial, but when phrased differently it can be quite mind-boggling, as we will now see.

Suppose that in the year 2000 we begin using a resource of some commodity. The annual extraction increases steadily and is doubled after every 10 years. (This doubling time corresponds to an increase by about 7 % annually, according to the 72 rule.) We also assume that the total resource that is available will be exhausted in the year 2100. Table 4.3 shows how much remains during the twenty-first century.

After half a century, 96.97 % remains. Two decades later, in 2070, there is still about 88 % left. With such a slow depletion it may seem that there is no reason for concern. But the reality is quite different. The resource that had been used for 70 years will be gone in just another 30 years.

Now suppose that in 2070 someone discovers another resource just as big as the first one. People might then say that since only about 12 % of the previously known resource had been used between 2000 to 2070, with the new finding there should be no fear of shortage for a very long time to come. And yet the new resource pushes forward the time to complete exhaustion by only one doubling period — from 2100 to 2110 in our example.

Can one construct a scheme for the use of a finite resource so that

Table 4.3. Depletion of a resource with exponentially increasing extraction

Year	*% remaining*	*Year*	*% remaining*
2010	99.90	2060	93.84
2020	99.71	2070	87.59
2030	99.32	2080	75.07
2040	98.53	2090	50.05
2050	96.97	2100	0

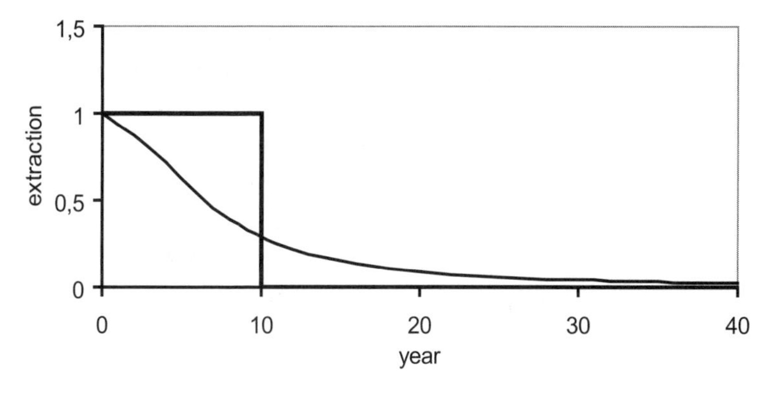

Fig. 4.1. Two scenarios for the extraction of a finite resource. One of them extends to infinite times.

it lasts forever? Figure 4.1 shows a case (thick line) where the available resource would last 10 years if the extraction is kept constant.[2] As an alternative (thin line) we can start with the same extraction the first year but then let it gradually decrease, under the constraint that the area under the two curves (thick and thin lines, respectively) taken to infinite times is the same. We could even allow the consumption to increase initially if this is followed by a tail with a much lower extraction. In fact, there are infinitely many ways to devise extraction rates such that a resource is never completely exhausted. This is the nearest we can come to sustainability when the resource itself cannot be replaced.

4.2 The Ubiquitous Straight Line

Whether female marathon runners will soon outrun men, laws of nature halt technical development, and increased flying cause many cancer deaths.

Dubious Extrapolations

In 1992 the prestigious journal *Nature* published a letter by Brian J. Whipp and Susan A. Ward entitled "Will Women Soon Outrun Men?"[3] The authors looked at graphs giving the average speed for men and women during the twentieth century until 1992 for the 200-, 400-, 800-, 1500-, and 10 000-m events and for the marathon. Data for women, which are available for a shorter period of time, showed a higher progression rate than for men. The conclusion was that the results "will be no different for men and women within the first half of the twenty-first century. . . . Beyond that time, current progression rates imply superior performance by women. The projected intersection for the marathon is 1998." Of course, this article was not intended to be taken as a serious forecast, but it can be used to illustrate the dangers in extrapolation of data.

Let us now take a new look at this type of analysis. From 1964 to 1983 the world record in women's marathon was improved 11 times, from 3 h 19 min to 2 h 25 min. Figure 4.2 shows these data. In this graph a best fit to the data points (in a root-mean-square sense) is extrapolated forward in time linearly. According to this extrapolation the projected record by the beginning of the twenty-first century would be below 1 h 40 min. This can be compared with the winner's time, 2 h 26 min 44 s, at the 2008 Olympics in Beijing. The winning time for men was 2 h 06 min 32 s.

Another interesting exercise is to extrapolate backward in time. If we use the graph in figure 4.2, the time to complete the marathon run would increase without limit. Instead, we therefore plot the inverse of the time (equivalent to a plot of the average speed) versus the year. A straight line is fitted in the region of the data points.

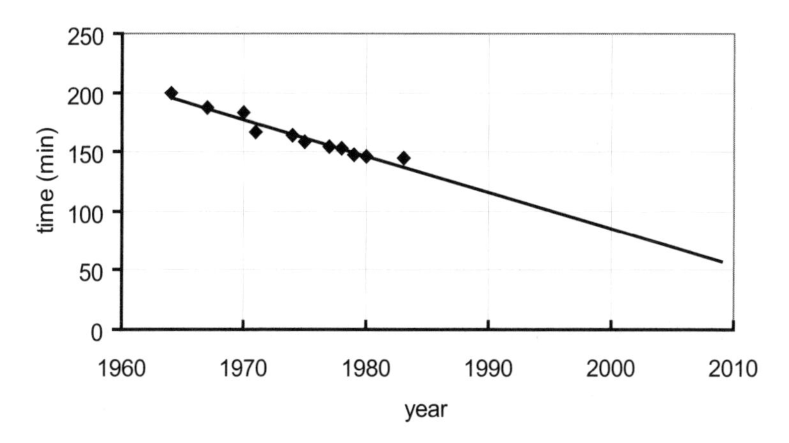

Fig. 4.2. Extrapolation of women's world records in the marathon from 1964 to 1983

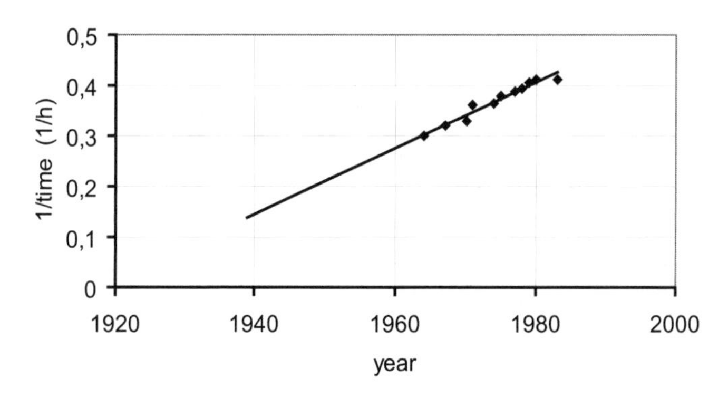

Fig. 4.3. The average speed for the women's world record in the marathon, expressed as the inverse total time, from 1964 to 1983

The result is shown in figure 4.3. Obviously an extrapolation gives zero speed in the early twentieth century, suggesting that the runner would then stand still. Earlier than that, the speed would be negative — in other words, the competitor is running backward. But we don't need to make such ridiculous comments to realize the dangers in extrapolations, forward or backward, when the data are limited to a narrow interval.

Moore's Law

In the *Electronics Magazine* issue of April 19, 1965, Gordon Moore wrote an article with the title "Cramming More Components onto Integrated Circuits." There the text reads: "The complexity for minimum component costs has increased at a rate of roughly a factor of two per year. . . . Certainly over the short term this rate can be expected to continue, if not to increase. Over the longer term, the rate of increase is a bit more uncertain, although there is no reason to believe it will not remain constant for at least 10 years."

The term "Moore's law" was coined five years later by Caltech professor Carver Mead. In 1975, Moore suggested that the doubling time would be *two* years. It is important to note that the prediction was about the complexity for minimum cost, not how many transistors there can be per square inch or the like. A colleague of Moore at Intel modified the prediction to allow for the improved performance of transistors and concluded that integrated circuits would double in *performance* every 18 months.

Since then there have been numerous discussions of what is called Moore's law, looking at various properties related to electronics. For instance, the hard disk capacity of PCs in gigabytes has increased by a factor of about 100 per decade. An analogous increase has been noted for the number of pixels per dollar in a digital camera, corresponding to a doubling about every 18 months. The power consumption of computer nodes shows a similar doubling time, and the computing performance per cost in dollars doubles at a somewhat slower rate. All such analyses plot the logarithm of a quantity versus time. If the result is a straight line, there is exponential growth with a certain doubling time.

It is not unusual to refer to Moore's law in descriptions of all kinds of technological development, or even worse, to *assume* that the "law" is actually obeyed in such cases—a generalization that Moore himself refutes. Raymond Kurzweil, an American inventor and author of many books on the future technological society, looked at computing machines and their performance expressed as the number of operations per second.[4] These machines ranged from the mechanical machines used in the 1890 US census, to computers based on relays, vacuum tubes, and today's transistors. There is an approximate exponential growth in performance during the first half of this period, and then a growth of about twice that rate during the second half, which is described by Moore's law. Could this progress continue?

Ultimately there will be physical limits on how small a transistor can be made, but perhaps that is not the correct way to view the problem. In all technical development one may ask if the laws of nature pose insurmountable restrictions. In some cases that is really so. For example, as we will see later in this chapter, there is a minimum energy requirement to synthesize ammonia from nitrogen and hydrogen. We will now look at two other cases that are completely different.

Although it sounds ridiculous, the Sun radiates like a black body. Physics tells us that all matter sends out radiation, called black-body radiation, which varies very rapidly with temperature. An iron bar at ambient temperature emits light that falls predominantly in the infrared region, which we cannot see. If the bar is "red hot" we can see the emitted light, but it is still not bright enough to be a useful light source. Biological evolution has given us eyes that are most sensitive in the range of light frequencies sent out by a body with the surface temperature of the Sun, about 6000 K. A body heated to that temperature would be a good artificial light source.

During the nineteenth century many inventors worked on incandescent light bulbs. The problem was to find a material for the filament that could last for a reasonably long time and be heated to high temperatures. Carbon offered the best compromise, but it had a reddish light because the filament temperature was rather low. The next material tried, around 1900, was a tungsten filament. Tungsten is the element with the highest melting temperature, about 3400 °C (6200 °F), which is still well below the temperature at the surface of the Sun. Thus, nature seemed to limit technical development in this area. Then fluorescent tubes were invented. In fluorescent bulbs the visible light does not originate from a hot body but from electrons that are accelerated in an electric field. Through the choice of coatings on the inner tube walls, the dominating frequency (the "color") of the light could be made more like that of the Sun. Today, still another light source is rapidly gaining use—the environmentally friendly light-emitting diode (LED).Thus, the real challenge is not to find a material with a high melting point: it is to provide light. That can be obtained in other ways than by black-body radiation from hot filaments.

The development of the telephone provides another example of how technical barriers can be overcome. Telephone communication over long distances once required wires with a very low electrical resistance. Copper was the metal of choice, but it was expensive. There are no other low-resistivity metals (except for silver and gold), so nature seemed to hamper the development of reasonably cheap telephone networks. The real problem, however, is not to send a current; it is to transfer a spoken message. That can be done with light in glass fibers. Glass, essentially made of sand, is available in unlimited amounts and at a reasonable price. We can also send the message with electromagnetic waves, as in mobile phone systems. The role of the telephone—to allow people to be in simultaneous contact—has not changed much since the days of Alexander Graham Bell, but the technical solutions have overcome the apparent limit set by the metals that nature gives us.

The purpose of technical development is to satisfy human needs—for instance, information storage, light, and communication. When a certain technology seems to be limited by the laws of nature, the problems may be circumvented through the use of an entirely different technical solution.

Low Radiation Level and Cancer

Extrapolation usually means that one wants to predict performance at longer times, higher loads, and so on. But sometimes one wants to extrapolate from a region where an effect is clearly documented to a region where the effect is extremely small. An important example is

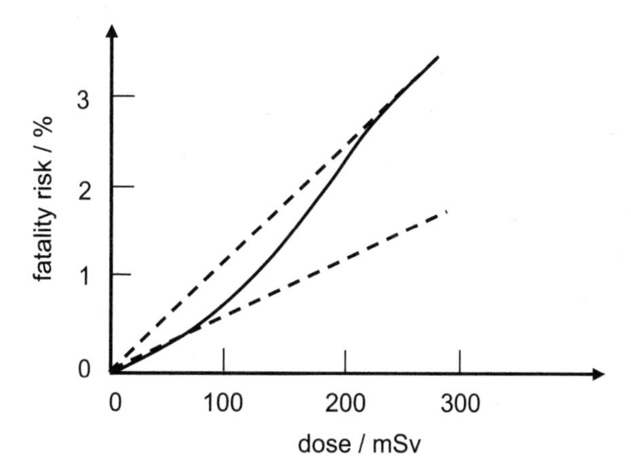

Fig. 4.4. The fatality risk according to the linear no-threshold (LNT) hypothesis (dashed straight lines) and a tentative interpolation (solid curved line) between the high-dose and low-dose LNT descriptions

the dose-response correlation for people subject to harmful agents. Actually, this is a kind of interpolation rather than extrapolation, since the response is assumed to be zero in the absence of exposure to the agent.

We know a lot about the risk of death due to radiation when the dose is high and of short duration. One obvious source of such information is from Hiroshima and Nagasaki. The problem is how to use these data when the radiation intensity is very low but the exposure time is very long. There are two fundamentally different approaches. In one approach we extrapolate linearly from the region of high doses down to the origin in the dose-response plot. In the other approach, linear extrapolation ends at a finite dose level (a threshold), below which the response is assumed to be zero. In 1955, the International Commission on Radiological Protection (ICRP) adopted the former approach, called the linear no-threshold (LNT) hypothesis. At high doses the mortality risk is known to be proportional to the dose (upper dashed extrapolation in fig. 4.4). For low doses, ICRP still assumed that the response is proportional to the dose, but now with a slope that is halved (lower dashed extrapola-

Table 4.4. Typical radiation doses

Source	*Dose* (mSv/year)
External background (highly variable)	1
Natural ^{40}K in body	0.4
Medical X-rays for broken leg	0.3
Air travel, return trip USA–Europe	0.05
Nuclear power, world average	0.0002

tion in fig. 4.4). The solid-line curve in the figure is an attempt in this book to interpolate between the linear regimes.

Here we are concerned with so-called ionizing radiation, in contrast to electromagnetic radiation such as light from the Sun, a fire, or a lamp, or radio waves from radio and TV signals. Sources of ionizing radiation occur naturally and cannot be avoided. Our bodies contain the radioactive isotope potassium-40 (^{40}K). Further, there is background radiation from radioactive materials in the Earth and in the form of cosmic radiation from space. The latter is damped in the atmosphere, but we receive a significant dose when we are flying. The chemical element radon is a gas which emits radiation. Radon emerges naturally from the ground, but with very large geographical variations. The most important man-made source of ionizing radiation is X-rays used in medical practice. In comparison with that, the radiation from nuclear power plants is negligible (see table 4.4). The unit for the dose equivalent in SI (Système International d'Unités) is sievert, denoted Sv. It is common to express the dose in mSv (millisievert). There is also a non-SI unit for dose, the rem, with 1 rem = 10 mSv.

The linear extrapolation down to zero dose in the LNT model is not based on scientific evidence. Instead, the hypothesis was introduced as a way to establish limit values for radiation. A dose of 4000–10 000 mSv causes acute illness and death within days. If the LNT model is strictly applied, it means that in a population of one million people the natural background radiation (about 1 mSV per year) may cause the death of several hundred humans per year. It is extremely difficult, not to say impossible, to verify such a prediction because we are all subject to the varying natural background radiation, and there are many common but complex reasons besides radiation for the development of cancer.

4.3 Scaling

On obesity and the size of fish, why Gulliver was given too much to eat and drink, and how long it takes to roast an ostrich.

Big and Small Fish

The mathematical formula used to calculate the body mass index (BMI) relates height and mass. This leads us to the field of allometry in biology — that is, how a property Q depends on the body mass, M, of an animal through the relation

$$Q = bM^a.$$

The word *allometry* comes from the Greek *alloios*, meaning "different," and refers to an exponent $a \neq 1$. (Note that $a = 1$ is an isometric relation, i.e., "by the same measure.") If there is an allometric relation, a plot of $\log Q$ versus $\log M$ yields a straight line, from which one can obtain the prefactor b and the exponent a, as we can illustrate with fish.

The simplest allometric relation to look for is that between typical size and mass. However, it is not always obvious what measurements should be chosen for size, particularly if we are comparing animals of very different shapes. Four-legged animals could be characterized by the maximum height from the ground when the animal is standing upright, or the height to the withers (as for horses and dogs), or the length from the nose to the beginning of the tail, and so on. Determining size within a single species is much simpler, since the individuals have similar shapes. Let's take pike as an example.

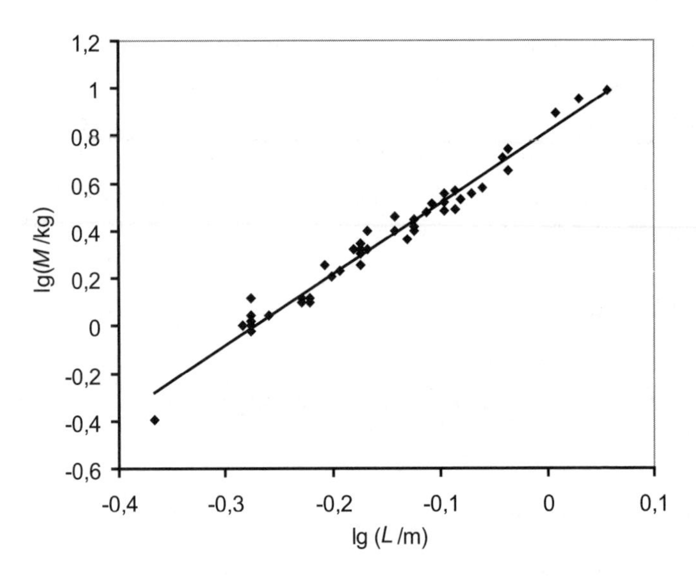

Fig. 4.5. The relation between mass, M (kg), and length, L (m), for pike

Figure 4.5 shows the masses of 43 pike (*Esox lucius*) in a plot of log$M$ versus log$L$, where M is the mass and L is the total length. The result is a straight line, showing that there is an allometric relation. The slope of the line gives the exponent, $b = 2.99$, so the mass is very accurately obtained from the cube law, which holds if all individuals are scaled in size without any change in shape. There is an "outlier" of very small mass in figure 4.5. That individual was found in the stomach of one of the larger pikes.

The result for pike seems to be at variance with the definition of BMI, where one takes the inverse square, rather than the cube, of the height. We return to this issue in section 7.4, What Is Your BMI?

Gulliver

Jonathan Swift's book *Gulliver's Travels* gives a familiar example of scaling. In the miniature land of Lilliput, Gulliver met people who were 6 inches tall, while Gulliver stood 6 feet. In the giants' land of Brobdingnag the inhabitants were a factor of 10 larger than Gulliver

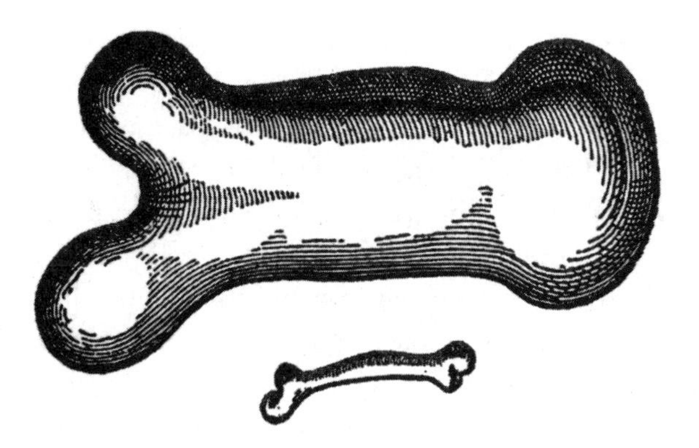

Fig. 4.6. The scaling of bone size. From Galileo Galilei, *Dialogues concerning Two New Sciences* (1638).

in all their dimensions. But even before Swift's book, Galileo Galilei had noted that similar animals (e.g., cats and lions) are not geometrically scaled by the same factor in all dimensions. A small animal has more slender legs. If all dimensions are increased by the same factor, say 10, the mass, and hence the weight, would increase by a factor of 1000. But the cross-section area of the legs and the bones, which are to carry the load, would increase by only a factor of 100 (see fig. 4.6). In *Dialogues Concerning Two New Sciences*, Galileo writes:

> I have sketched a bone whose natural length has been increased three times and whose thickness has been multiplied until, for a correspondingly large animal, it would perform the same function which the small bone performs for its small animal. . . . If one wishes to maintain in a great giant the same proportion of limb as that found in an ordinary man he must either find a harder and stronger material for the bones, or he must admit a diminution of strength.[5]

Gulliver found the voices of the giants to be unusual, but there is more to say about this. The characteristic frequencies of an animal's vocalizations are well known to decrease with the animal's size. (A British film comedy from 1959 with Peter Sellers and Jean Seberg, on the theme of nuclear threats, had the title *The Mouse That Roared*.) As we speak, our tongues move back and forth. If the giants in Brobdingnag tried to utter words at the same pace as Gulliver, their tongues should also move back and forth just as rapidly. According to Swift's description of the giants' size, the distance a giant's tongue is moved would be larger by a factor of 10. Therefore, the acceleration in the movement of the tongue would also be larger by the same factor. Further, the mass of the tongue is larger by a factor of 1000. According to Newton's law, force is acceleration times mass. It follows that the force that is required to move a tongue this large must increase by a factor of 10 000. But the number of strands of muscles providing that force has increased only as the cross-section area of the tongue root—in other words, by a factor of 100. Moving the tongue of a giant as fast as Gulliver's tongue would be like waving a wet towel.

Gulliver's Travels (published in 1726) has four parts. In the first two of them, Gulliver visits Lilliput and Brobdingnag. In the other two parts he travels to the floating island of Laputa and to a land of intelligent horses. The work is neither a children's book nor a novel, but a bitter satire attacking the British society of the time, in particular, the academic establishment. Any criticism of Swift's book on the grounds that it has many physical and physiological "errors" is of course pointless.

Roasting a Turkey

Roasting a turkey is not something one does every day, and the birds can vary a lot in size. No wonder there are many recipes that give tables of roasting times based on the mass of the turkey. The Food Safety and Inspection Service of the US Department of Agriculture (USDA) says that roasting a turkey of 12 pounds requires about 3 hours, and a turkey twice as large requires about 5 hours, when the oven temperature is 325 °F (160 °C) and the turkey is not stuffed. Note that the cooking time increases less rapidly than the increase in mass.

How does this recommendation stand up to a comparison with a more scientific approach to the cooking time? Let us start with an object (for instance, a turkey) that has a low and uniform temperature throughout, and is put it in an environment of high temperature. Heat propagates into the object like a "heat front." That is a rather vague concept, but a mathematical model would tell us that the penetration depth of the heat increases as the square root of the time elapsed since the object was put in the hot environment. This model works well for most foodstuffs — fish, meat, potatoes, and so on. Of course, the propagation of heat depends on the conduction properties of the "material," but the message can always be expressed as follows: Double the thickness to the center, and the cooking time will increase by a factor of 4. A more general mathematical formulation is

> *The cooking time varies as the square of a typical length characterizing the object that is cooked.*

If all turkeys have the same shape (a reasonable approximation), their linear dimensions scale with mass, M, as $M^{1/3}$. If we combine this with the rule that the cooking time, t_{cook}, varies as the square of the characteristic linear size, we get

$$t_{\text{cook}} = kM^{2/3},$$

where k is a proportionality constant. Let us fix k so that t_{cook} agrees with the recommended cooking time of 4¼ hours for a medium-sized turkey of 18 pounds. Figure 4.7 shows a plot of the cooking time versus the size of the turkey, according to the formula above. The recommendations by the USDA are published as an interval in cooking time for each listed interval in mass. The filled circles in figure 4.7 give the values at the two end points of such intervals.

We see that the mathematical formula and the USDA recommendation give approximately the same result in the range of typical sizes. This is not surprising since we fitted the equation to the recommended data for a particular turkey mass (18 lb). The agreement at this point is perfect by construction, but also the other recom-

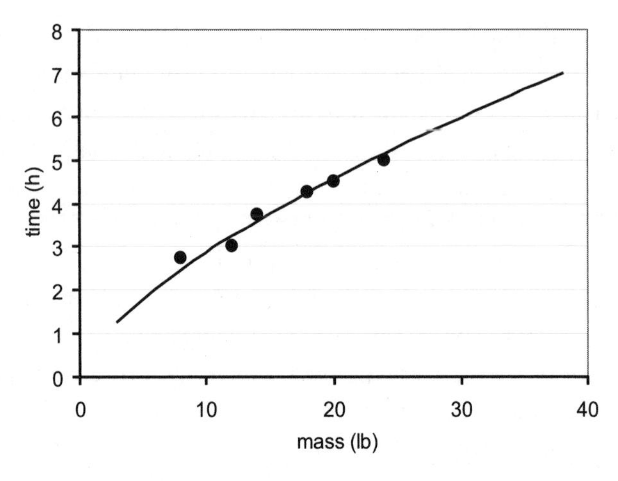

Fig. 4.7. Time required to roast a turkey, as a function of its mass. The points are recommended times, and the curve is a fit with a theoretical model.

mended cooking times are well reproduced by the model. Therefore, we may have some confidence in an extrapolation of the theory to much larger sizes. The mass of an ostrich can be about 130 kg (290 lb). If we had an oven that was large enough, it would take about 27 hours to roast that bird, provided that heat penetrates in similar ways in turkey and ostrich — a reasonable assumption.

4.4 Looking Ahead

Why academic studies can be a questionable investment, why economists refer to the second derivative, and why the lynx population in Canada varies periodically.

The Law of Diminishing Returns

A farmer's sowing is a kind of investment, in the hope that there will be more grains returned. The first barrel of wheat grains distributed over an empty field yields a certain harvest. The harvest from the next barrel planted in the same field probably yields somewhat less. At least, it will not be more in return than from the first barrel. As more and more seed is added to the field, the return per barrel gradually shrinks. One may speak of diminishing returns.

We get another twist on the same problem if we put one price (a cost) on the seed and another price (a profit) on the harvest. In the beginning, each additional barrel of seed costs less than the value of the increased harvest, but eventually one will come to a point where planting more actually decreases the profit.

There are numerous examples in our daily lives of these two aspects of diminishing returns. Consider, for instance, the value of academic studies and take it to the extreme. We may think of a student who wants to get a broad education, rather than going deeply into a special field. Courses in economy are followed by the study of several languages and then perhaps history and so on. This is done in the hope of becoming more attractive on the job market. The cost is considerable, both in time and money. Taking some academic courses probably is better than no studies at all. But soon each additional course adds very little to the student's attractiveness on the job market.

To take another example: an increased amount of training improves an individual's performance in sports and other areas like playing chess or an instrument or speaking a foreign language. The effect is largest in the beginning. Each additional hour of training gives a gradually smaller improvement in return. This is recognized in the points given to the events in decathlon and heptathlon, as we saw in the section on track and field in chapter 2. It is much easier to improve your results in high jump by 5 cm if you are a mediocre athlete than if you belong to the world elite. Therefore this improvement is worth fewer points if you are well below the elite level. In other words, the diminishing return as measured by the result in centimeters is compensated for by giving more points per centimeter improvement at the elite level.

In the industrial world it usually costs more in investment for the next improved generation of a product. There may even be a theoretical limit, which can never be reached but only approached asymptotically. One example is the energy cost to produce ammonia,

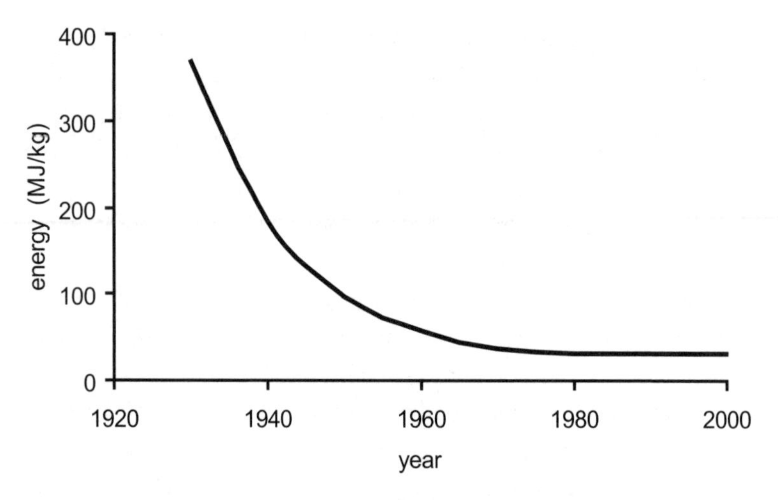

Fig. 4.8. Energy required for the industrial production of ammonia. Graph redrawn from Vaclav Smil, *Energies* (Cambridge, MA: MIT Press, 1999).

which has the chemical formula NH_3. The German chemists Fritz Haber and Carl Bosch achieved an industrial breakthrough in the early 1900s when they developed the Haber-Bosch method to synthesize ammonia from nitrogen and hydrogen gas. Fundamental thermodynamics gives a minimum energy cost for the synthesis. There are always losses in the industrial process, and to reduce them further usually becomes increasingly difficult. Figure 4.8 shows how the energy required to synthesize ammonia has decreased with time and now flattens out asymptotically toward about twice the theoretical limit.

The Sign of the Second Derivative

Table 4.5 shows the hypothetical sale of a certain product per quarter over 3½ years. With such a steeply rising growth, it may seem attractive to invest money in the manufacturing company. But the numbers deserve a closer look. The prospects are not as good as they first appear to be when one considers only the sales figures in the first

Table 4.5. Hypothetical development of the sales of a product

Time	*Sales ($ × 1000)*	*Growth (%)*	*Change in growth (%)*
Year 1, Q 1	47		
2	60	13	
3	76	16	3
4	95	19	4
Year 2, Q 1	119	24	4
2	148	29	5
3	182	34	6
4	223	40	6
Year 3, Q 1	269	46	6
2	321	52	6
3	378	57	5
4	438	60	4
Year 4, Q 1	500	62	2
2	562	62	0

column. The second column gives the change in sales between consecutive quarters. There is still a steady growth. In fact, the growth rate has never been as large as during the last two quarters. However, the third column tells another story. It gives the "change in growth"—that is, how much the data in the preceding column vary from one quarter to the next. That quantity has dropped during the last year and is likely to become negative. This is what both mathematicians and economists would call a change in the second derivative. The first derivative is represented by data in column three (growth in sales) and the second derivative by data in column four.

The development in table 4.5 is characteristic of what is called a sigmoid, or S-shaped, growth. A common mathematical form of a sigmoid curve is the logistic function or logistic curve. It is shown in figure 4.9, with the data points from the table marked.[6] (In

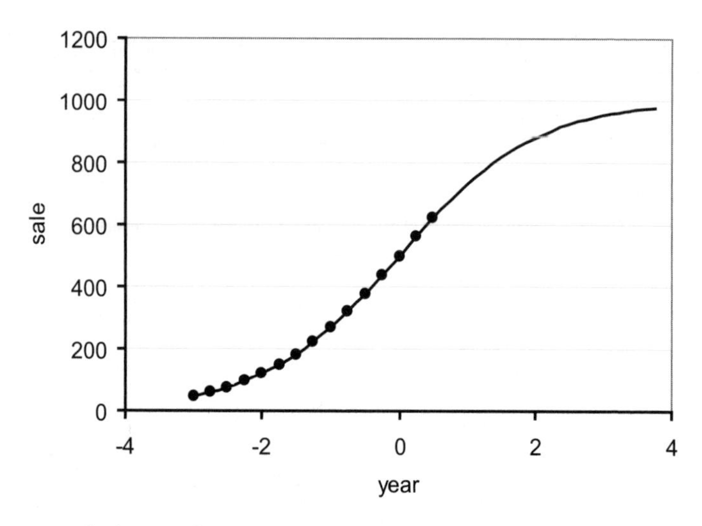

Fig. 4.9. The logistic function (curve) and the sales data (points) from table 4.5

fact, those hypothetical data in our example were generated from the logistic function, so the exact agreement has no significance.) The *slope* of the curve at a certain point is given by the first derivate, and the *curvature* (bending) by the second derivative. We get a clear warning about the coming saturation when the curvature changes from being convex (curved upwards in the figure) to concave. The important message for those who want to look into the future without being fooled by the growth rate alone is the following rule:

The growth rate has its largest value when the second derivative changes sign, indicating a subsequent decline in growth.

That is why we look for a change in the sign of the second derivative.

There are several other types of sigmoid functions, having mathematical forms that are somewhat more complicated than the logistic curve.[7] They find application in a large number of processes in addition to commerce—for instance, in the modeling of tumor growth, chemical reactions, and ecological systems.

Lynx and Hare

When prey is abundant, predators get more food, reproduce easily, and increase in number. As a consequence, the prey will be under pressure, and its population will decrease. With less food available, the predators struggle more to survive and therefore decrease in number. Then the prey get a chance to become more abundant again. This is the idealized predator-prey interdependence. The populations of predator and prey rise and fall in a periodic pattern. There are, of course, many external factors which could also affect the variations in population, such as weather, diseases and parasites, wildfires, limited size of the ecosystem, and human interference like logging. If such external influences are small, we expect that peaks of prey are correlated with peaks of predator, which come somewhat later in time.

The classical example of predator-prey interdependence is that of the lynx and snowshoe hare in Canada. Figure 4.10 shows part of the famous data for the populations of snowshoe hare (thousands) and lynx (hundreds), based on the records of fur trade kept by the Hudson's Bay Company. The data points in our example are marked for every second year, starting in 1845. We see that the peaks come with a periodicity of about 10 years. Other related data show that this periodicity is quite stable, with nine strong peaks between 1845 and 1935.

There is a mathematical model, the Lotka-Volterra equation, which reproduces a periodic pattern like that in figure 4.10. However, one must ask if there are not other explanations than a simple predator-prey mechanism. Since the Canada lynx–snowshoe hare case has such a prominent place in the scientific literature, it has been the subject of many detailed studies.[8] Basically, the population size depends on the reproduction rate and the mortality. The mortality

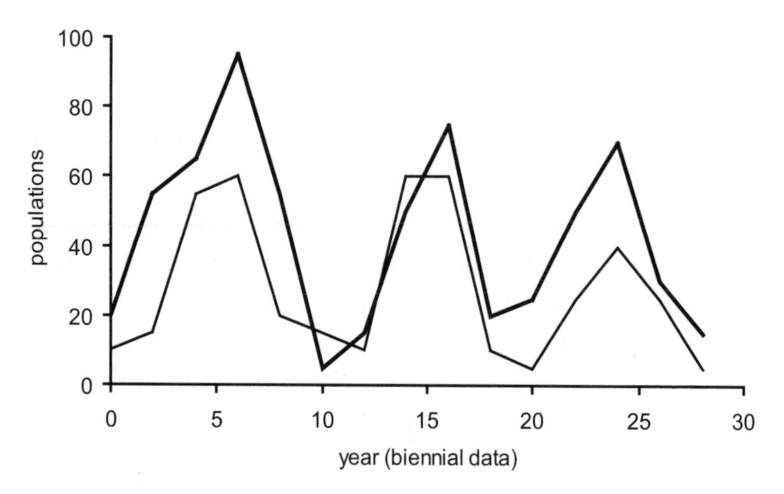

Fig. 4.10. Biennial data for the populations of snowshoe hare (thousands; thick line) and lynx (hundreds; thin line), based on the fur trade records kept by the Hudson's Bay Company

rate of the snowshoe hare is almost entirely driven by predation, but there is an interaction between predation and food supply. Predators other than the lynx play a secondary role. During the peak and decline phases of the hare population, lynxes may kill more hares per day than they need. When the hare population has crashed, many lynxes starve to death. In particular, the female has difficulty feeding both herself and her kittens, who succumb.

It is generally accepted that the lynx-hare relation can be well explained by the predator-prey mechanism, in spite of several complicating factors. But that leaves one question to be answered: Why is the periodicity about ten years? One could think of possible external driving forces. For instance, it has been noticed that the sunspot cycle is about 11 years, and there have been speculations that this cycle implies related periodicities in the ecosystem. However, no such connection has been verified. The lengths of the reproductive generations of lynx and hare set obvious time scales, but there seems to be no clear explanation why this would lead to the 10-year intervals in population peaks.

There are cyclic phenomena, such as the tide, which are driven by an external force but also many which are "self-driven." Some people would put the sequence of economic booms and recessions in the latter category. In humans we may think of the heartbeat and the menstrual cycle. The "chemical clock," in which the color of a mixture changes periodically, is another cyclic phenomenon that is sometimes shown in chemistry classes and science centers.[9]

5 Models

5.1 What Are the Chances?

On how many typos remain after proofreading, how bad it is to lose an arm if you have already lost a leg, and why travel time can be shorter if a road is closed.

Proofreading

Proofreading is a tedious task, and you might think that a single reading would do, if it is careful enough. Unfortunately, that is seldom true. It can be a very frustrating experience to go through the text a second time and discover many mistakes that you did not see the first time. A better alternative is for two people to do the proofreading independently. There will be errors that both proofreaders notice but also errors that are discovered by only one of them. With this information one can estimate how many have still gone unnoticed. For instance, suppose that the two proofreaders, A and B, have a 50 % chance of detecting an error. If there are 100 errors in a text, A and B each will find approximately 50, of which 25 are noted by both. In all, 75 different errors are detected, and 25 remain.

We now construct a general model.[1] Let a text have M errors, which A and B notice with probabilities p_A and p_B. The product $p_A p_B$ gives the probability that a particular error is detected by both A and B. On average, they find Mp_A and Mp_B errors, respectively, of which $Mp_A p_B$ are seen by both of them. The number M can be trivially expressed as an algebraic identity:

$$M = \frac{(Mp_A)(Mp_B)}{(Mp_A p_B)} \; .$$

Because the three expressions enclosed by parenthesis on the right-hand side are known from the result of the two independent proofreadings, we can obtain M. The number of remaining errors is the total number M minus those already detected:

$$M - (Mp_A + Mp_B - Mp_A p_B).$$

The term $-Mp_A p_B$ is a correction to avoid double-counting of the errors that were discovered by both proofreaders.

As an example, suppose that proofreader A found 30 errors. Proofreader B found 20 errors, and 15 of them were noticed also by A. Then an estimate of the total number of errors in the text is

$$M = \frac{30 \times 20}{15} = 40.$$

The number of errors found by either A, B, or both of them is $30 + 20 - 15 = 35$, and from the formula above we expect $40 - 35 = 5$ to remain undetected. The model is based on probability arguments, so its result must be interpreted with some caution, but it still gives a good estimate. Note that we don't need to know p_A and p_B (i.e., how good the proofreaders are).

Losing a Leg

What is worse—losing your right leg or losing two fingers? That depends. Most people would say that losing a leg is worse. But for a professional trumpet player, that may not be so. Insurance companies must address such questions when they give compensation for disability or impairment due to accidents. According to the American Medical Association, *impairment* is defined as any loss or abnormality of psychological, physiological, or anatomical structure or function. *Disability* is an alteration of an individual's capacity to meet personal, social, or occupational demands because of impairment. An individual can be impaired—for instance, wheelchair-bound—but nevertheless be able to perform her profession; she is therefore not in this respect disabled. Likewise, a very minor impair-

Table 5.1. Severity of impairment values used by Swedish insurance companies

Impairment	%
Loss of left leg	35
Loss of index finger	7
Loss of thumb	19
Blind, one eye	14
Completely blind	68

ment can make an individual completely disabled in performing his professional tasks.

It is not easy to set a scale from 0 to 100 % where 100 corresponds to complete impairment or disability. Physicians who are asked to evaluate an individual case could easily come to very different results. Therefore, insurance companies and authorities use tables which assign values to blindness in one eye, the loss of the right-hand thumb, and so on. The numbers in such tables can refer to impairment or disability, and assign weight to profession and age. Insurance companies also have lists that focus only on the loss of a certain anatomical structure or function. Table 5.1 shows an example. Here, the ratings are given without regard to age, profession, responsibilities, hobbies, and so forth.

It could happen that an accident leads to multiple losses. Obviously, one cannot just add the percentages, since the result may exceed 100 %. The combined rating R_{combined} in the case of two losses is calculated as

$$\frac{R_{\text{combined}}}{100} = \frac{R_1}{100} + \frac{R_2}{100}\left(1 - \frac{R_1}{100}\right),$$

where R_1 and R_2 are the ratings (in percentages) of losses 1 and 2. For example, if we use the data from table 5.1, the simultaneous loss of the left leg (35 %) and one thumb (19 %) is considered as a percentage loss amounting to $35 + 19(1 - 0.35) \approx 47$. The generalization to three or more separate and independent losses is straightforward.

The method is called *cumulative weighting*. It has many applications outside the field of medicine, as is illustrated by the following example. Suppose that transportation of supplies to a famine region can only be by truck. Originally there is a certain full capacity, rated as 100 %. If half of the trucks have an engine breakdown, transport capacity is reduced by 50 %. Similarly, if heavy rain means that the speed is reduced to half of its normal value, while all the trucks can operate, the transport capacity is also reduced by 50 %. If now half of the truck engines break down, and the roads are simultaneously damaged by rain, the resulting transport capacity is, of course, not reduced to 0 %, but to 25 %.

We can extend the example to illustrate one more important aspect. Suppose that in normal operation there is enough fuel to run all the trucks. Due to some circumstances the available fuel has been limited and suffices for only half of the trucks, thus reducing the transport capacity by 50 %. If that fuel shortage coincides with a heavy rainfall, the transport capacity is reduced to 25 %. But if half of the engines then also break down, this circumstance will not further reduce the capacity. Only half of the trucks could be used anyway because of the fuel shortage. Similarly, the simultaneous loss of certain functions of the leg, like the knee, the ankle, and so on should never be rated as more severe than losing the entire leg.

Sunday Traffic

Every Sunday afternoon people return from the coastal resort A to their hometown D. There are three possible routes for the cars: ABD, ACD, and ABCD, as shown in figure 5.1. The roads AC, BC, and BD are of high quality, and require 2 h, 1/4 h, and 2 h in driving time, respectively. The roads AB and CD are rather short but easily get congested. The driving times are $(1 + p)$ hours and $(1 + q)$ hours,

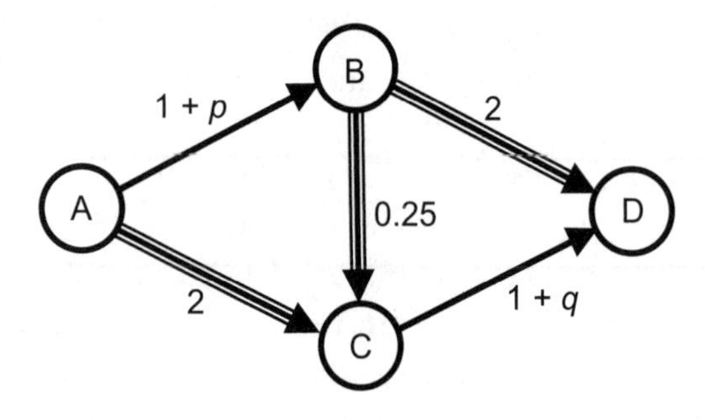

Fig. 5.1. Travel times for alternative routes from A to D

respectively, where p and q give the fractions of all returning cars that take those routes ($0 \leq p \leq 1$, $0 \leq q \leq 1$). Everyone wants to return home as quickly as possible.

With very light traffic, it would be best to take the road ABCD, giving the travel time 2¼ h. This is the fastest route until $p = q = 0.75$. At that point the travel time has increased to 3¾ h, the same as the travel time along the previously slower routes ABD and ACD. Now some people will go directly from A to C, but since that would cause more congestion in link CD, some of those arriving from A to B will prefer to take the slow road BD. This pattern is repeated every weekend, and gradually people learn that when $p = q = 0.75$ there is nothing they can do to shorten their own travel time. It is 3¾ h for all three alternative routes.

One day a bridge on the link BC is so severely damaged that the road is closed for a very long time. Now there are only two routes, ABD and ACD. They are equivalent, except that one starts, and the other ends, with a long road. Soon the traffic becomes equally distributed on the two routes, with $p = q = 0.5$. To their surprise people now find that they all have a total travel time of 3½ h, in other words, 1/4 h shorter than when they could use the road BC.

When the bridge is repaired, and traffic resumes on the link BC, the travel time increases again, contrary to what common sense would tell us. The origin of this paradoxical result is that everyone acts in an egoistic way. The situation with $p = q = 0.75$ is an equilibrium in the sense that for no individual is there a better alternative. If the cross link is closed, individuals will take whichever leg (ABD or ACD) has the lowest traffic. With everyone acting like that, traffic will be equally divided between the two legs. The travel time could be reduced to 3½ h if everyone obeyed a collective recommendation not to use the cross link. But then some individuals would find that they can reduce their own travel time even further, to 3¼ h, if they first choose the AB link and then continue on BC rather than on BD.

The situation with $p = q = 0.75$ is an example of a Nash equilibrium. It is named after John Forbes Nash, Jr., who received the 1994 Nobel Prize in Economic Sciences for his work on the concept of noncooperative games in game theory. (The Prize in Economic Sciences, established in 1969, is not one of the prizes named in Alfred Nobel's will but is awarded "in memory of Alfred Nobel.") Nash's life became known to a wide audience through the Hollywood movie *A Beautiful Mind*, which won four Oscars. There is a famous bar scene in the movie about how to approach a group of women when one of them is considered more attractive than the others. Nash suddenly leaves the bar because, according to the film story, he got the idea that later resulted in his Nobel Prize.

5.2 Seeking the Optimum

Why some people argue that a lower tax rate gives higher collected tax, how you should run on the beach to rescue a drowning child, and why it is not necessary to customize a golf club head to obtain optimum mass.

Tax Rates and the Autobahn

The name of the economist Arthur Laffer is associated with a curve that purports to describe the total collected income tax as a function of the tax rate. There is a legend about how Laffer sketched the curve on a napkin at a meeting in 1974, but Laffer himself recalls that he had often used the curve in classes and discussions, and that it has roots as far back as the Muslim scholar Ibn Khaldun in the 14th century, and much later to John Maynard Keynes. Figure 5.2 shows the characteristic shape of the curve. The vertical axis gives the collected tax revenue as a function of the income tax rate. At rates above a certain value, the revenue decreases. The implication is that to the right of the maximum in the curve, a reduced tax rate will increase the incentive for people to work and therefore increase the total collected tax.

This is a highly idealized scenario that has been much criticized. It is trivial to understand that the curve starts at zero, but it is not evident that it should also end at zero for a 100 % tax rate. If the curve shows the marginal tax, it is only the top part of the income that is taxed according to the graph. One could imagine a society where there is a highest allowed income. Work beyond that is in some sense voluntary, and the society keeps all the revenue. In the real world there can also be several ways of tax evasion. Further, above a certain income level benefits can be given in nonmonetary forms that are not subject to tax.

In spite of all the criticisms of the Laffer curve when it is taken too literally as a model of taxation, it nicely illustrates a mode of thinking that is highly relevant far outside the field of economics — in physical sciences, engineering, biology, and elsewhere. The basic idea is that at the extreme ends of a certain phenomenon, the "effect" may be small or zero, for more or less obvious reasons. Between these ends there is a maximum, or optimum. Moving somewhat to either side of such an extremum does not significantly change the value of the effect. It is a merely a *second-order correction.*

As an application of this mode of thinking, consider traffic flow, for instance on a German autobahn (highway), where there are no speed limits. If the speed of the cars is zero, the flow rate (the number of vehicles passing per hour) is of course zero. One might naively think that the higher the speed, the larger the flow rate. Double speed would give double flow rate. But stopping distance increases as the square of the speed. Therefore, the distance between cars must increase more rapidly than the speed itself if road safety is main-

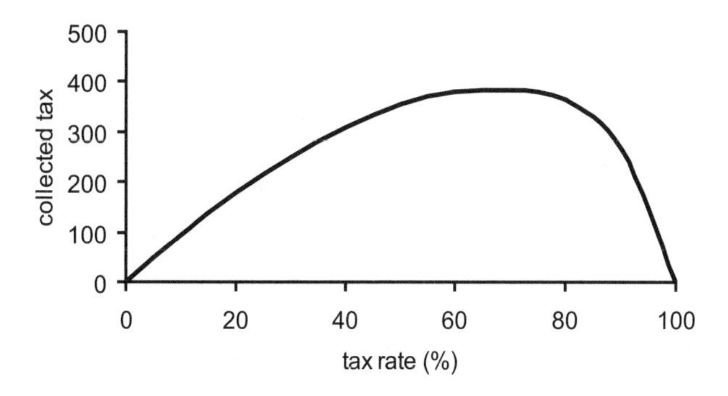

Fig. 5.2. The Laffer curve (schematic)

tained. It turns out that in a broad range of intermediate densities (around 30 vehicles per kilometer) and a speed of about 50 km/h (30 mile/h), the flow rate has a maximum around 1500 vehicles per hour. This is the ideal behavior, but our model ignores the familiar instabilities that can cause a traffic standstill for no apparent reason. It also ignores reckless driving at high speed.

Biology is full of examples where the Laffer type of curve provides a crude basis for the thinking. The male peacock displays his feathers in order to attract the females. If the feathers are small, he will not look impressive, but if they are too large, there will be obvious disadvantages. As another example, some migrating birds return each spring from Africa to northern Europe to breed. If they come early, they may secure a good territory, but then they also suffer the risk of succumbing to a last spell of cold winter weather. The peacock's feather size and the arrival date of the migrating birds can vary within a range of almost equally good alternatives, in agreement with the shape of the Laffer curve.

Running to the Rescue

When introductory physics books discuss the refraction of light at the boundary between air and water, or between air and glass, it is not unusual to make an analogy to the following kind of problem. You are standing at point A on a beach, when you see a screaming

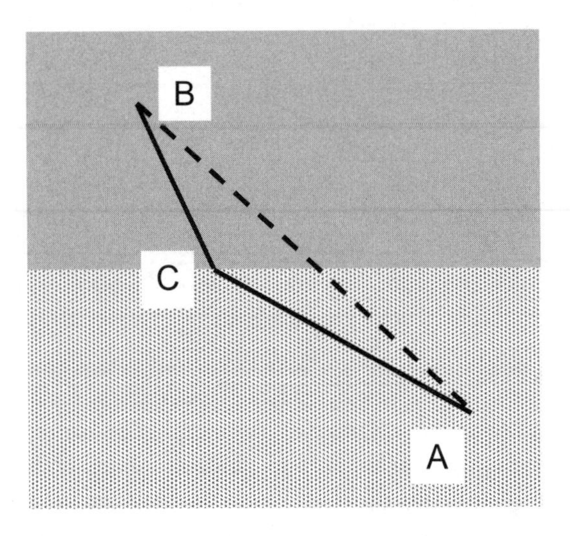

Fig. 5.3. Two paths from A on the beach to B in the water

child sinking below the water at point B (fig. 5.3). Quickly you decide to rush to save the child, but because it is much faster to run on the beach than to cover the remaining path in the water, you should not take a straight path between A and B. The problem is to choose the location of point C that minimizes the total time to the child.

Those familiar with physics know that the problem is closely related to Snell's refraction law in optics, but that is not the point of interest right now. Instead we address the following question: How important is it that we choose the correct location for C? In other words, does it matter much if C lies somewhat to the left or to the right of the optimum location? With some effort we may work out a detailed mathematical solution, but it is better to do some thinking. The point C represents an extremum—the path that gives the shortest time to get from A to B. Let us sketch the time as a function of the position along the shoreline where you enter the water (fig. 5.4). The curve has a minimum at C. In the immediate vicinity of that point, the time varies very little. Therefore, it is not important if you hit the shoreline right at the point C, or somewhat to the left or to the right.

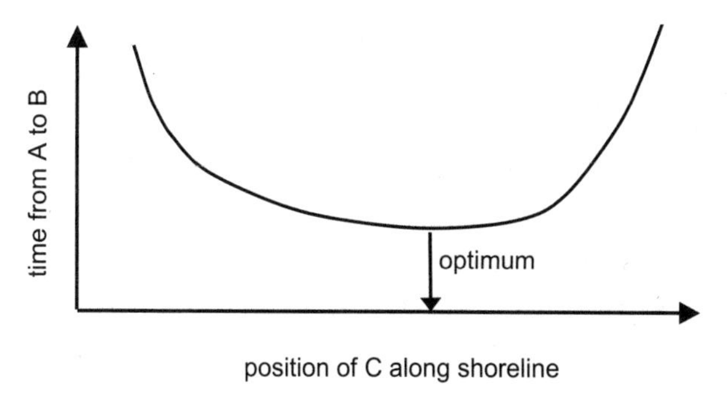

Fig. 5.4. Freehand sketch of the typical variation in total time from A to B in fig. 5.3, as a function of the location of C along the shoreline

In the language of mathematics, the change in time is a second-order effect. When you dash off to save the child, your intuition will certainly tell you this, without any thoughts about the mathematics.

Second-order effects can also be viewed as corrections to a correction. When a spring is pulled, its length increases. The elongation is doubled if the force is doubled. This is the famous Hooke's law, named after the British scientist Robert Hooke (1635–1703). He wanted to keep his discovery secret for a while, without losing future credit, and therefore announced the rule as the anagram *ceiinosssttuu*. It was later revealed to give the Latin phrase *Ut tensio sic vis* ("as the extension, so the force"). Hooke's law represents the first-order correction to the length when a force is applied. It is accurate enough in most engineering applications, but in some cases one must allow for further corrections—that is, second-order terms in the theory.

Selecting the Best Golf Club

Golf rules specify in great detail the properties of the ball but leave more freedom when it comes to the clubs. For instance, the ball must not have a diameter less than 1.680 inch (42.67 mm), and its mass must not exceed 1.620 ounce (0.04593 kg). Now let us seek the optimum mass of the club head to get the maximum initial speed of the ball in a drive.

It is a good approximation to model the shot as the collision between a club head of mass M and a ball of mass m, thus ignoring the shaft of the club. Before we invoke mathematics, it is illuminating to look at two extreme cases. If the mass of the club head is zero ($M = 0$), the ball will remain at rest. At the other extreme, if the club head is infinitely heavy, we could not swing the club. The ball would also remain at rest in that case. Somewhere between these limits is an optimum mass for the club head, with a value that may depend somewhat on the player. But small variations around the optimum mass will give only a second-order correction to the speed of the ball. Therefore, it is unnecessary to customize the club head mass for each player.

The following description of the shot uses some simple physics, but even without that knowledge it should be easy to follow the arguments. In a drive, the club head approaches the ball with a certain speed, v_0. But only the *relative* motion is important. Therefore, we could also view it as if the club head is at rest, with the ball coming toward it with the speed v_0. Let us first imagine that the club head is infinitely heavy. The incoming ball will bounce back with the speed v_1, as if it had hit a rigid wall. If the collision is completely elastic, the speed of the ball will not change, i.e., $v_1 = v_0$. But that is an idealization. With a real ball, the rebound speed is reduced by a factor e, called the restitution factor, and becomes

$$v_1 = ev_0.$$

Golf balls usually have $e \approx 0.8$. However, we are interested in the speed, v, of the ball relative to the ground, and not relative to the club head. Thus, we must add the speed, v_0, of the club head relative to the ground to get the desired speed of the ball as

$$v = v_0 + ev_0.$$

In the next step of our argument, the club head mass is no longer assumed to be infinite but has a value M. This step involves a rather elementary application of momentum conservation in physics, but

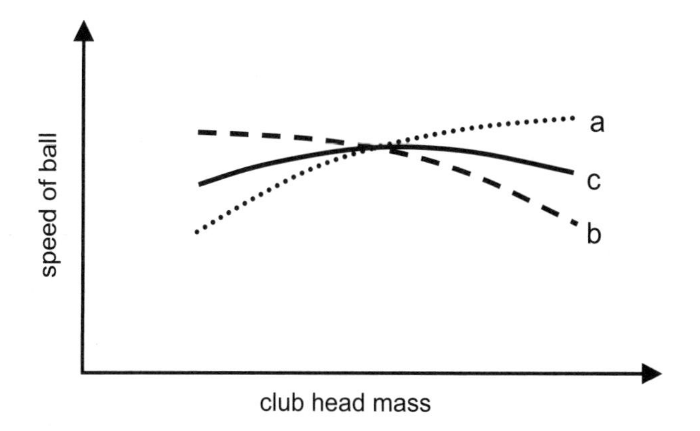

Fig. 5.5. A schematic representation of the final speed of the golf ball as a function of the mass of the club head if we (a) only take into account the collision between club head and ball, (b) only take into account that a heavy club is more difficult to swing, and (c) combine the two effects

we will skip those details and just give the final result for the speed of the ball relative to the ground, which is

$$v = \frac{M}{M+m}\,(1+e)v_0.$$

Typically, the mass, M, of the club head is 0.3 kg—i.e., $m/M \approx 0.15$. If M is changed by $\pm 10\,\%$, it will change the speed of the ball only by $\pm 1.5\,\%$. An increase in the club head mass would give a slightly higher speed for the ball if it were not for the fact that a heavy club is more difficult to swing. These two opposing effects are shown schematically in figure 5.5.

Our model is extremely simple. For instance, it does not account for the properties of the club shaft or how the hands connect to the shaft. However, it captures the main features and gives a surprisingly good description. In particular, the weak dependence of the speed on the mass of the club head is also obtained in more elaborate models, as long as the mass of the club head is much larger than the mass of the ball. In the spirit of the Laffer curve, we could have come to that conclusion without any mathematical analysis.

5.3 Focus on the Essential

How a spherical mouse explains why Gulliver was fed too much, how a baked potato has something to do with the age of the Earth, and why a sober party may still become noisy.

How Small Can a Mouse Be?

Mammals in a cold environment can maintain a constant body temperature if the heat generated by metabolism is large enough to compensate for the heat loss to the surroundings. Because that requires a certain body size, a very small mouse may freeze to death. Metabolic heat is very roughly proportional to the mass of the animal (but see the refinement below), and heat loss is roughly proportional to the animal's surface area. As a drastic simplification, let an animal (for instance, a mouse) have the shape of a sphere with radius R (fig. 5.6). Its area varies as R^2 and its volume as R^3. If R decreases by a factor of 10, our simple model implies that the heat loss decreases by a factor of 100 but the heat production goes down by as much as a factor of 1000. This scaling behavior holds not only for a sphere, but for any specified shape of a body, and is usually referred to as the *square-cube* (or cube-square) *law*. Since the density of the flesh does not vary much, the mass will also scale as the cube of the linear size.

There are many studies of animals in which the metabolic rate, the body surface area, or some other property, Q, of interest, is plotted versus the animal's mass, M, on a log-log plot. If the allometric equation

$$Q = bM^a$$

is exact, one obtains a straight line with a slope given by the exponent a (cf. sec. 4.3, Big and Small Fish). Data for the metabolic rate of mammals and birds, with M varying by more than 5 orders of

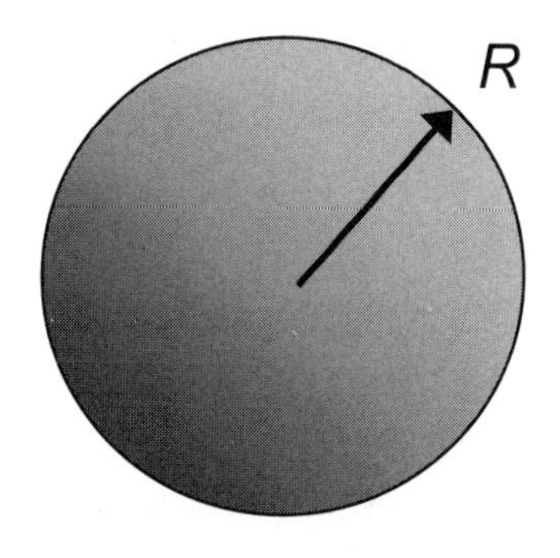

Fig. 5.6. A scientist may model a mouse as a sphere with radius R.

magnitude from mouse to elephant, fit this equation very well, with $a = 0.75$. If there were a direct proportionality between the body mass and the generated heat, we would have $a = 1$, but a smaller animal has a somewhat increased metabolic rate, which helps to keep it warm. Similarly, the body surface area of vertebrates fits the same type of relation with $a = 0.63$, rather than $a = 2/3 = 0.67$ that follows from a strict square-cube law.

Returning to the smallest size of a mouse, we could take data for the metabolic rate and an estimate of the heat loss per body area that is exposed to the surroundings, work out the details, and find that the smallest mice should have a body mass of a few grams (a tenth of an ounce).

Gulliver's Travels also provides a good illustration of scaling and metabolism. Gulliver's ration in Lilliput was very generous, as is clear from this quotation:

> The Reader may please to observe, that in the last Article for the Recovery of my Liberty the Emperor stipulates to allow me a Quantity of Meat and Drink sufficient for the support of 1728 Lilliputians. Some time after, asking a Friend at Court how they came to fix on that determinate Number; he told me that his Majesty's Mathematicians, having taken the Height of my body by the help of a Quadrant, and finding it to exceed theirs in the Proportion of Twelve to One, they concluded from the Similarity of their Bodies, that mine must contain at least 1728 of theirs, and consequently would require as much Food as was necessary to support

that number of Lilliputians. By which the Reader may conceive an Idea of the Ingenuity of that People, as well as the prudent and exact Oeconomy of so great a Prince.

Gulliver could not eat and drink that much. In line with the fanciful descriptions in *Gulliver's Travels* we rely on the square-cube law and see what would happen in the reversed case, in which the Lilliputians got 1/1728 of Gulliver's food ration. If they were physiologically similar to Gulliver, their metabolism would provide an amount of heat to the body that was smaller by a factor of $12^3 = 1728$. The cooling of the body decreases only by a factor of $12^2 = 144$, if we assume it to be proportional to the body area. The Lilliputians could not maintain the normal body temperature of 37 °C and would freeze to death.

The Age of the Earth

If the text in the Old Testament in the Bible is taken literally, the Earth was created about 6000 years ago. Modern science, on the other hand, finds its age to be about 4 500 million years. In the nineteenth century, the age of the Earth and our solar system was a matter of much debate. Geologists assumed that the geological processes we can witness today had "always" been effective. Then they could estimate, for instance, how long a time it would take for erosion to change the surface of the Earth. The result was that the Earth must be at least many hundreds of millions of years old. Charles Darwin found such estimates to be in accord with his own ideas about evolution as a slow process. Then in 1862, the British physicist William Thomson published what is arguably the most famous and most debated physical model calculation of the nineteenth century.[2] Thomson, who later became Lord Kelvin, was one of the scientific giants of his time, and his name has been immortalized in the temperature unit of the SI system.

To understand the essence of Kelvin's model calculation of the age of the Earth, we may first think of a baked potato. When the potato has been baked, we take it out of the oven and put it on a plate to cool. The cooling is a rather slow process because the stuff the potato is made of is not a very good conductor of heat. After some time, the parts close to the skin have cooled significantly, but the interior is still quite hot. The temperature in the baked potato, which was first almost uniform from the surface and inwards, now varies with the distance to the potato skin.

Kelvin constructed his model in analogy to our baked potato. He knew that the Earth's temperature increases by about 1 °C for every 30 m depth into the ground (1 °F per 50 ft). Then he assumed that the Earth had once been molten rock, which required a temperature of almost 4000 °C (7000 °F). From mathematical modelling of the cooling of a sphere, he concluded that the Earth was between 20 million and 400 million years old, with preference given to values close to the lower limit. This stirred up a lot of controversy, because on one hand it was enormously longer than what the Bible said, but it was also much shorter than what the geologists could accept. Today we know that Kelvin underestimated the Earth's age by about two orders of magnitude.

So what went wrong? With modern computing facilities it is not more than a homework problem in an engineering course to repeat the numerical calculations in Kelvin's model. In fact, they are so simple that this is not where Kelvin erred. The relevance of a model rests on its assumptions. Kelvin assumed that there was no other heat source inside the Earth once it had been formed. Further, he did not account for the possibility that there is convection in a fluid core at great depths. Both of these assumptions were incorrect.

Radioactivity, discovered by Henri Becquerel in 1896, can liberate large amounts of heat. It is widely thought that radioactivity, which Kelvin of course had no chance to anticipate, explains his too short age estimation. If there is an internal heat source, Earth will take longer to cool. However, it is now known that radioactivity alone does not suffice to explain the present temperature gradient at the Earth's surface. Convective currents are another very important aspect neglected in Kelvin's model.

It would be wrong to criticize Kelvin for the large error in his estimate. After all, in spite of being a religious man, he took a scientific approach that led him to results in strong conflict with the official Christian doctrine, and he did so on the basis of the best physics knowledge of his day. Kelvin's model of the age of the Earth is famous in the history of science as a reminder that the assumptions underlying a model may turn out to be incorrect in the light of new facts. Science is a matter of continuous improvement in our understanding of nature. Knowledge is provisional, but that is not to say that all old knowledge will eventually turn out to be wrong. Einstein's theory of relativity, and Schrödinger's equation in quantum mechanics, do not invalidate Newton's equations as a basis for the modelling of, for instance, a car crash.

A Loud Party

Who has not been to a party which has become so loud that normal conversation is impossible? The cause may not be excessive drinking. Even formal and sober parties can be loud, due to simple principles of acoustics.

The modelling goes like this.[3] The party takes place in a large ballroom. Guests arrive one by one. Having learned all the rules of social behavior, they act accordingly and form small groups. Within each group there is always an ongoing conversation, but only one person at a time speaks in a group. The listeners in that group can understand what is said if the background noise is not too loud compared with the conversation they are involved in. Engineers would express this requirement in the form of a quantity, the *signal-to-noise ratio*, S/N, which must be large enough.

The strength of the signal, S, reaching the listener's ear is directly proportional to the power output, P, of the speaker. Further, according to a simple geometrical argument, the signal decays as the square of the distance, d, between the speaker and the listener in a group. We can write this as

$$S = k_1 \frac{P}{d^2},$$

where k_1 is a proportionality constant that is not of immediate interest.

The noise, N, received by a listener is caused by the conversations in all the other groups. In each of them there is a speaker, whose power output is also P. The groups are assumed to be randomly distributed over the ballroom. The noise is the result of sound being reflected by walls, floor, ceiling, groups of other party guests, and so on. We let N be proportional to the number of people at the party, n:

$$N = k_2 P n.$$

Here k_2 is another proportionality factor. The signal-to-noise ratio becomes

$$\frac{S}{N} = \frac{k_1 P / d^2}{k_2 P n} \sim \frac{1}{n d^2}.$$

The symbol $\sim$ means "varies as" or "is proportional to."

Now let more people arrive, so that n increases. As a consequence, the signal-to-noise ratio, S/N, decreases. The guests compensate for this by coming closer together (making d smaller), so that S/N still allows a comprehensible conversation.[4] But with more people arriving, one reaches a critical distance $d = d_c$ which is the smallest tolerable. Obviously d_c varies with the local social conventions, for instance, often being smaller in South America than in northern Europe. After this minimum distance is attained, the speakers find no other way to make themselves heard within a group than to speak louder—in other words, to increase their power output, P. But as we have just seen, that does not help because P appears both in the numerator and the denominator in our model for S/N and cancels out in the final expression. Every speaker at the party now desperately increases P with no other result than making the party noisy.

The modelling of a phenomenon can have different objectives. One extreme case is to formulate models that give very precise *quantitative* descriptions—for instance, how much a bridge sags under a certain load. The opposite is a model that serves the purpose only of identifying an important aspect or mechanism, to give a *qualitative* description without any attempt to provide accurate numerical results. In both cases a key issue is whether the model is *robust*—that is, whether conclusions drawn from it depend crucially on assumptions and the value of input parameters.

Our description of the loud party is only qualitative but still quite robust. It focuses on the instability connected with the fact that the distance between speaker and listener cannot be made arbitrarily small. We assumed that the noise is proportional to the number of guests. However, all that is needed for our argument is the obvious fact that the background noise increases with the number of guests; a strict proportionality is not necessary. Similarly, the assumption that S varies as $1/d^2$ is not crucial. It suffices that the signal increases if the speaker and the listener come closer together. No doubt our model has identified an important aspect of why a party may become loud.[5]

5.4 A Law or a Model?

Why Ohm's law is not a law, why one should not waste time on inventing a perpetuum mobile, and why experiments always have the last word in physics.

Ohm's Law Is Not a Law

Physics is full of named laws that are expressed as simple mathematical relations: Kepler's laws of planetary motion, Boyle's law for gases, Snell's refraction law in optics, and so on. One of the best-known laws from schoolbooks is Ohm's law:

$$U = IR.$$

But Ohm's law is not a true law of nature. It is just an approximation, although a very good one in many practical applications like the electrical wiring in a house. It relates the electrical current I, flowing through a resistor with resistance R, to the potential drop U

over the resistor. In the SI system we would measure I in ampere (not to be written "Ampere" or "ampère"), R in ohm and U in volt. Of these units, ampere stands out as more fundamental because it is one of the seven base units in SI. (The other six units are kilogram for mass, meter for length, second for time, kelvin for temperature, mole for amount of substance, and candela for luminous intensity.) The resistance R can be viewed as a proportionality factor between the electrical potential (voltage) U and the current I. It depends on the size and shape of the resistor and on the material it is made of. However, if the material is, for instance, a semiconductor, we no longer use Ohm's law. Engineers then talk of non-ohmic behavior.

Let us now turn to the three other laws mentioned above. We will find that they are of different character. Boyle's law implies that the gas pressure is doubled when a gas is compressed to half its volume without any change in temperature. This is an approximation and is sometimes completely misleading. For instance, a gas may condense to a liquid on compression. Boyle's law is a special case of the ideal gas law. The latter, as is obvious from its name, is an idealization of the real behavior of gases. It can be derived as a mathematical model under a number of approximations (idealizations) and is not a law of nature.

Kepler's laws describe the motion of the planets around the Sun — for instance, how the period for one full revolution (a "year") scales with the distance to the Sun. These laws can be obtained as consequences of Newton's equations of motion. Today, Kepler's laws are more like exam problems in a physics course, but they were certainly not so for Kepler, who died 12 years before Newton was born.

Snell's law of refraction describes the fastest path of light between points on the opposite sides of a boundary between two media, given the speed of light in each of these media. (An analogous problem is discussed in sec. 5.2, Running to Rescue, above.) The result belongs to mathematics just as much as to physics.

Many relations in science and engineering relate three quantities A, B, and C to each other, for instance, in equations like $A = BC$ or

$A = B/C$. It is not unusual for students who take introductory physics courses to get the impression that memorizing such "formulas" is what physics is about. One relation they are likely to encounter is

$$\rho = \frac{M}{V} .$$

It gives the (mean) density, ρ of a material if its mass, M, and volume, V, are known. This relation is not a law: it is a definition. As such, it is exact, provided that the mass and the volume are well defined in the case one considers.

In physics and related fields there are many mathematical relations that are commonly called laws, or principles. In most cases, however, they are not truly basic laws or principles but can be derived from more fundamental starting points. In the absence of a "theory for everything" there are nevertheless some relations that are so fundamental that they should be called laws of physics. Among them are Newton's equations in mechanics and the Schrödinger equation in quantum mechanics, although these are only approximations to a description that includes relativistic effects. There are also fundamental principles, like the first and second law of thermodynamics, which are justly referred to as laws of nature. But in spite of what has now been said, we should certainly continue to use phrases like Ohm's law.

A Mad Pursuit

Astrology and perpetuum mobiles have no place in modern science. They are two examples of pseudoscience, although they usually attract different kinds of people. While astrology belongs to the realm of beliefs, those obsessed with the construction of a perpetuum mobile — a perpetual motion device — are often very honestly trying to apply the principles of physics. (This, of course, excludes those who consciously commit fraud.) Looking back at the history of perpetuum mobile one encounters the names of several famous scien-

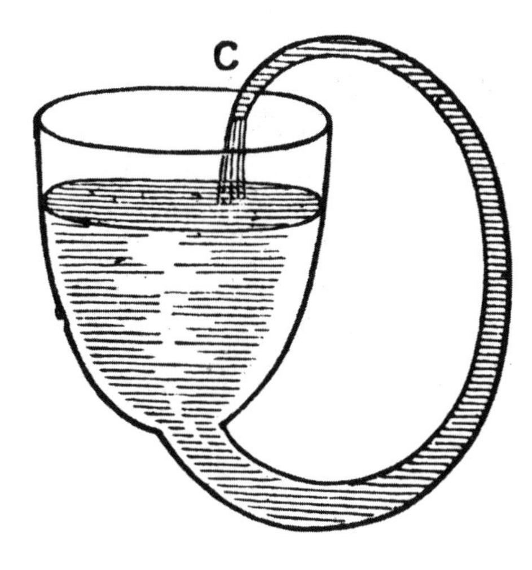

Fig. 5.7. A perpetual motion device attributed to Dennis Papin and usually called Boyle's flask

tists who lived at a time when there was not yet a good understanding of the concept of energy.

Figure 5.7 shows a device constructed by Dennis Papin in 1685. It is usually referred to as Boyle's flask, with reference to Robert Boyle (1627–91), who is known for his law for gases. The perpetuum mobile action in the flask was thought to be caused by surface tension. Often, surface *tension* problems are easier to understand if one instead uses the concept of surface *energy*. Tension is expressed with the SI unit N/m (newton per meter), but since $N/m = (N \cdot m)/m^2 = J/m^2$, the numerical value of surface tension expressed as force per length is the same as the value of energy per area. In Boyle's flask the liquid may rise inside the narrow tube because the total energy of the system is lowered if the *air*-glass boundary inside the tube is replaced by the *liquid*-glass boundary. Theoretically, the surface properties could be such that the liquid reaches the mouth of the tube. But then there is no more glass surface to cover with liquid. The formation of a protruding liquid drop would increase the area of the liquid-air boundary and therefore increase the total energy. That cannot happen spontaneously.

If Boyle's flask really worked, one could use the kinetic energy of the liquid drops that continuously fall from the mouth of the flask. That is in violation of the first law of thermodynamics, which says that energy cannot be created or destroyed: it can only be converted between different forms like potential and kinetic mechanical energy, electrical energy and heat. This is often called the energy principle. (It is extended through the famous relation $E = mc^2$ in relativity — an aspect that we ignore here.)

Beginning at least as early as in the thirteenth century, there have been numerous attempts to find a machine that would circumvent the first law of thermodynamics. New ideas pop up even today, as many professors of physics can testify. Usually it is not too difficult to spot the mistake. This is not true with constructions violating the second law of thermodynamics. That law can be formulated in several seemingly different ways, which are in fact equivalent. One version says that heat cannot spontaneously flow from a cold place to a warm place — an intuitively reasonable restriction. It can also be formulated in a much more subtle way, as a theoretical limit on the efficiency with which heat can be converted to mechanical energy. For instance, only a certain fraction of the heat generated in the car's combustion engine can be converted to the mechanical energy that drives the car forward.

Attempts to overcome the second law of thermodynamics can contain obvious errors, but they may also be so sophisticated that their analysis becomes the subject of research papers. No serious scientist believes that the law does not hold. However, it can be very instructive to pin down exactly where there is a crucial step in the construction. After many such analyses, the second law of thermodynamics has become one of the strongest pillars in the foundation of physics.

Staff at patent offices are among those who believe that the first and second laws of thermodynamics cannot be violated, but sometimes the inventor is so clever in disguising that a claim is actually a perpetuum mobile that a patent has been granted. The UK and US patent offices have the following policies:

UK PATENT OFFICE:

Processes or articles alleged to operate in such a manner which is clearly contrary to well-established physical laws, such as perpetual motion machines, are regarded as not having industrial application.

US PATENT AND TRADEMARK OFFICE:

With the exception of cases involving perpetual motion, a model is not ordinarily required by the Office to demonstrate the operability of a device.

The following excerpt from the book *The Nature of the Physical World* (1927) by the British astrophysicist Sir Arthur Stanley Eddington (1882–1944), gives an almost century-old view of the laws of physics—a view that is still valid today:

> The law that entropy always increases, holds, I think, the supreme position among the laws of Nature. If someone points out that your pet theory of the universe is in disagreement with Maxwell's equations—then so much worse for Maxwell's equations. If it is found to be contradicted by observation—well, these experimentalists do bungle sometimes. But if your theory is found to be against the second law of thermodynamics I can give you no hope; there is nothing for it but to collapse in deepest humiliation.

Is Coulomb's Law Exact?

Two electric charges attract each other with a force that varies as the inverse square of their mutual distance. An analogous dependence on distance governs the gravitational force between two masses. The inverse square law for the force between two electric charges is called Coulomb's law, although the British clergyman and scientist Joseph Priestley (1733–1804) suggested, almost two decades before the Frenchman Charles Augustus Coulomb (1736–1806), that the interaction between two charges follows the same $1/r^2$ law that Newton had given for gravitation.

But how sure can one be that the force varies exactly as the inverse *square* of the distance, r—that is, as the power r^{-2}? Perhaps the -2 value of the exponent is just a convenient and accurate approximation, with the true value being better approximated by (for instance) -1.99998 or -2.0003. Intuitively, that seems highly unlikely. But physics is an experimental science, and in principle the value of the exponent must be obtained from measurements.

Let us write the exponent as $-2 + \delta$, where δ is a small positive or negative number. Experiments in 1772 by the British scientist Henry Cavendish showed that $|\delta| < 0.02$. The Scottish scientist James Clerk Maxwell found, about 100 years later, that $|\delta| < 5 \times 10^{-5}$. A modern experiment from 1971 gave $\delta = (2.7 \pm 3.1) \times 10^{-16}$. It is tempting to assume that one may theoretically show δ to be exactly 0, but theory alone cannot give a definite proof of physical laws. However, theory can provide a framework, or a network of knowledge, such that different physical phenomena are strongly linked together. If a particular "law" or theoretical description is modified, it will have implications for other aspects of physics. The history of science shows several examples of how a discovery affects what has been regarded as a law of nature. For instance, until radioactivity was discovered in 1896, it was thought that the chemical elements are stable and cannot be transformed into other elements. And before relativity theory in 1905, and the relation $E = mc^2$, it was thought that mass is always exactly conserved.

Light is a form of electromagnetic radiation. Thus, it may not come as a surprise that the electrostatic forces, expressed by Coulomb's law, can be described as mediated by light quanta (photons). In such a theory, we arrive at the result that $\delta = 0$ if the photon is a "particle" with no mass (its rest mass being zero).[6]

From modern astrophysical measurements it has been inferred that the r^{-2} law is a very good approximation, at least up to distances $r << 10^{24}$ m. This may be compared with the distance between the Earth and the Sun, 1.5×10^{11} m. Conversely, experiments at the atomic level, in which alpha particles (electrically charged helium nuclei) are scattered against thin foils, have verified that Coulomb's law holds exceedingly well down to a distance of about 10^{-13} m, or 1000 times smaller than the diameter of an atom. We may therefore conclude that no deviation from Coulomb's law with the exponent -2 has been detected. It holds with very high accuracy from distances several orders of magnitude smaller than the diameter of an atom, up to distances larger than the size of our galaxy.

You might wonder how it has been possible to measure a force with such an extreme accuracy. A direct measurement of the variation of force with distance will have an appreciable uncertainty. Instead, we can look for consequences of an exact r^{-2} law. For instance, if the law is exact, it implies that there is no potential difference between two points inside a charged so-called Faraday's cage.[7] When no potential difference is detected within the uncertainty of the measurement, it exemplifies a *null experiment*—an experiment that shows the absence of an effect. Another well-known null experiment was the attempt by Michelson and Morley in 1881 to relate the velocity of light to the velocity of the light source through an assumed ether.

6
The Real World

6.1 Plausible, but Not Correct

Why riding a bicycle has nothing to do with gyroscopic effects, a thickened church window nothing to do with the flow of glass under gravity, and the bathtub vortex nothing to do with the rotation of the Earth.

The Unridable Bicycle

A common demonstration in physics classes and science centers uses a bicycle wheel with handles on both sides of the axis. A student or visitor is asked to grab the handles and tilt the wheel when it is spinning. It is not as easy as you might think, and the wheel moves in a different direction than anticipated. Sometimes this exercise is performed with the subject sitting on a swivel chair, with the result that the chair starts to rotate as the bicycle wheel is turned. All this is meant to illustrate the properties of a gyroscope—in particular, that a gyroscope tries to maintain the direction of its axis of rotation.

If a hoop or, even better, a car wheel is let free to roll down an incline, it continues its path in a very steady motion. Small obstacles are negotiated without the wheel falling over. There is an obvious inherent stability caused by the rotation.

Given these experiences it is no wonder that so many people believe it is the gyroscopic effect from the spinning wheels that makes it so easy to maintain one's balance on an ordinary bicycle. This is probably one of the most common misconceptions in the application of the principles of physics to the experiences of daily life.

What else could there be to keep the bicycle upright? One obvious candidate is centrifugal force.[1] As we lean the bicycle into a turn, the front wheel turns in such a way as to steer into the direction that the bicycle leans. The bicycle tends to move in a circular path, with a centrifugal force now acting outwards, to prevent the bicycle from falling over. We can think of how cyclists lean when they turn a corner. This stabilization seems to be the reason why we can easily keep our balance, even without holding the handlebars. Could it be something that is inherent in the bicycle construction itself? Theoretical arguments about forces, their magnitudes and directions, may sound convincing, but there is no better judge than a stringent experimental test, and that has been carried out in great detail. The British chemist David Jones got so intrigued by bicycle physics that he decided to build an "unridable" bicycle.[2] His clever idea was to find out what would make a bicycle unstable, rather than following the trail of others and asking why it is stable.

Jones first investigated the possible role of a gyroscope effect by mounting an extra wheel on the front fork and arranging it so that it could spin opposite to the rotation of the normal front wheel without touching the ground. This did not make the bicycle difficult to ride, even when the cyclist was not holding the handlebars. This unexpected result prompted two other experiments. When a hoop was given a counter-rotating inner wheel, it could not be sent down a slope without rapidly falling to the ground. Likewise, when the bike was left to run down a hill without a rider, and with the extra wheel in a counter-spin, the bicycle also soon fell over. But a normal bicycle, without the extra wheel, did not collapse when sent down the slope. It seemed that the gyroscopic action was present in the riderless bike but was not important when someone rode it and made it heavier. But that is not the end of the story.

After many experiments it became clear how crucial the steering geometry is for the ability of the bicycle to stabilize itself when it starts to fall over. The only really unridable bicycle had its front wheel moved forward relative to the normal construction. A horizontal extension was mounted to the end of the fork so that the wheel could be fixed as in figure 6.1. What it is in the combination of a human and a bicycle that makes the bicycle so easy to steer and

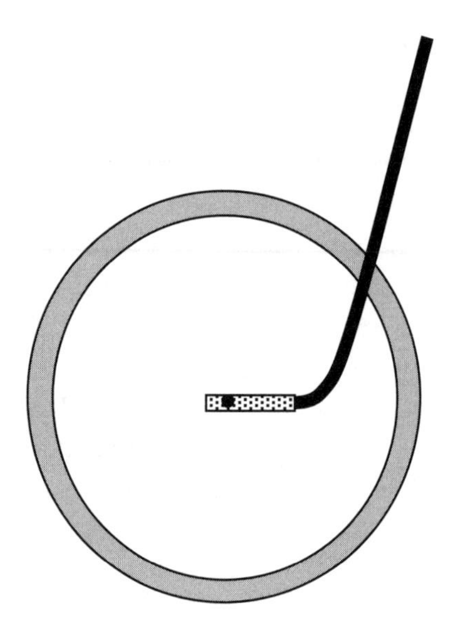

Fig. 6.1. A bicycle configuration that moves forward the contact point with the ground and therefore makes the bicycle more difficult to ride

keep balanced remains a complicated matter, but at least it is not a question of gyroscope stabilization.[3] Note, however, that a motor bike is something quite different. Its wheels are much heavier and are spinning much faster. Consequently, gyroscopic effects are essential.

Church Windows and Lead Roofs

It is a fact that old window glass panes have uneven thickness. There is also a widespread urban legend that they tend to have thicker bottoms because glass behaves like a viscous material. The gravity forces on a vertical sheet of glass are not very large, but if they act for many centuries, there can be visible effects — so the argument goes.

Old window panes are indeed often thicker at the bottom. Why is this so, then, if it is not because the glass flows? The most likely explanation has to do with how window glass was made. In one method, a cylinder of glass was blown, cut open, and flattened out on a hot surface. Then the panes were cut out, and they often had uneven thickness. One can imagine that it became customary to put the thickest end down when the panes were mounted in the windows. In another method, glass was spun to form a circular plate of uneven thickness. Modern panes are made with the float glass method, in which molten glass is spread onto a bath of molten tin. That gives a very uniform thickness.

It is understandable how the misconception about church windows has arisen. When glass is heated so much that it radiates light with an orange color, it behaves like a sticky liquid. Something that flows readily is said to have a low viscosity. The viscosity of glass increases tremendously as it is cooled, being perhaps 10 to 15 orders of magnitude larger at room temperature than at the temperatures at which it is handled by glass blowers. Such a large viscosity does not have much meaning as a measure of flow, and the material behaves as if it were solid.

It was not until the end of the twentieth century that we fully understood how wrong the legend about church windows was. The models used to describe the flow of glass require a deep understanding of physical mechanisms on the scale of individual atoms, accurate information about model parameters, and fast computers. When all that became available, it showed that a noticeable flow of window glass in churches is not possible.

The properties of glass may be compared with those of lead metal. It is well known that lead may deform in a way that can be described as a flow. Lead water pipes, mounted horizontally on the walls of old houses in England, have sagged under the combination of weight and time. When the National Cathedral in Washington, DC, got a new lead roof in 1919, it was not long before the roof showed severe signs of flow.[4] This was at first surprising, since lead roofs had been used without problems for centuries in Britain. One explanation could be the higher temperatures in Washington, but that was not the only reason. It was also the high price of silver and gold that, indirectly, caused the problem.

In earlier centuries, lead usually contained not only elements like antimony and bismuth but also small amounts of silver and gold. Then, there was no practical way to extract the precious metals, but by the early 1900s such methods were commercially feasible. In that process, lead also lost its content of antimony and bismuth. The new cheap lead was much purer than previously, but this turned out to be a disadvantage. The "impurities" are not a nuisance. On the contrary, they must be there to give lead a sufficient hardness and resistance to flow. Once this was realized, manufacturers found that alloying lead with 6 % antimony gave lead the desired properties. It is for the same reason that gold in jewelry is not pure (i.e., 24 karat). A ring of pure gold would be too soft; thus, 18 karat (i.e., 75 % gold) is a common composition.

The difference in flow properties between glass and lead at ambient temperature gives another striking argument against the old idea about the thickness of glass windows in churches. When glass is mounted in a lead caming, one would expect that the lead, rather than the glass, would flow. Indeed, there are such examples. (Apart from flow there is also an effect caused by the different thermal expansion of glass and lead.) The flow of glass panes in churches makes a nice story. Alas, it is just a legend.

The Bathtub Vortex

It is widely thought that when you pull the plug in the bathtub, the vortex rotates in one particular direction in the Northern Hemisphere and in the opposite direction in the Southern Hemisphere. If the experiment is done in an apartment building, there is a good chance that the vortices in the various bathtubs in that building will indeed rotate predominantly in one direction. Those who have taken advanced science courses may refer to the Coriolis force as the cause of the phenomenon. However, the bathtub vortex is a very complicated phenomenon. It is certainly the Coriolis effect that gives rise to the well-known clockwise rotations of the ocean currents and wind patterns in the Northern Hemisphere and the counterclockwise rotation in the Southern Hemisphere, but this effect is much too weak to govern the flow in an ordinary bathtub. The story about the bathtub vortex is just an urban legend.

However, in a *very* accurate experiment one may indeed see the result of Coriolis forces in vessels as small as a bathtub. At least three such experiments, performed in Massachusetts, England, and Australia, have been reported in scientific journals.[5] Of course, the experimenters were not primarily motivated by a need to check a theory but because of the folklore connected with the issue. The care by which the experiments were carried out is obvious from the following details in one of the cases.[6] A cylindrical vessel, about 2 m in diameter, was made of plywood, with a plastic and wooden cover and a small slit through which the vortex could be observed. The experiment was performed in a windowless basement, where a thermostat kept the temperature constant to within 20 ± 1 °C. All these precautions were to avoid currents caused by temperature gradients in the water. When the vessel was filled, the water was intentionally given a rotation opposite to that which was anticipated by the Coriolis effect; then the researchers waited 18 hours to let the initial rotation decay. When the plug was pulled out of the hole, which was placed in the center of the vessel, the vortex had the direction predicted by theory.

But what is the anticipated direction? Scientists often talk about a "parcel" of water or air to describe the motion of a small portion of a fluid or a gas. In the Northern Hemisphere, the Coriolis force affects a moving parcel of water or air so that it deviates to the right and hence has a clockwise rotation. It is then tempting to think that if one looks down from above on the bathtub vortex, it will have the same direction as the ocean currents and winds that are often depicted on a world map. But that is wrong! For an idealized bathtub experiment, the theoretical prediction is in the opposite direction. Look at figure 6.2. Water flows toward the outlet. Due to the Coriolis force, the flow is curved to the right (in the Northern Hemisphere). Therefore it will first "miss" the hole, but it is then "pulled back" toward the hole in a circular counterclockwise motion. It is the same as in the circulation of storms when viewed from above.

In a real bathtub, there will be currents from filling the tub that decay very slowly. Further, the shape of the tub and the outlet can

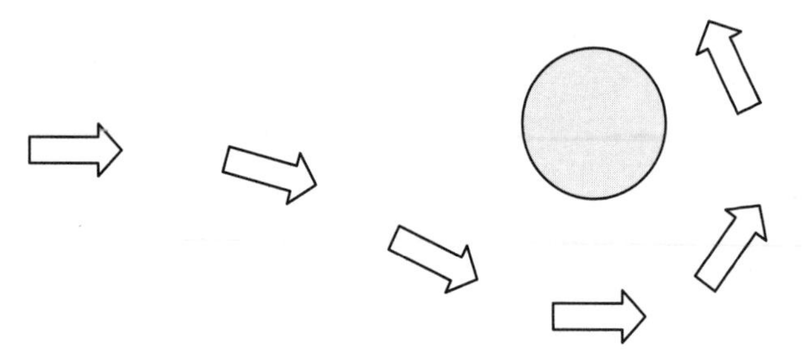

Fig. 6.2. Schematic of the motion of water toward the drain in the bathtub

have a significant influence on how a vortex is created. There is a good chance that all the tubs in an apartment building are very similar, which explains why one vortex direction could dominate in that particular building. This does not prove that the Coriolis forces are at work. Although such effects are present, they are much too small to be of any relevance.

What about the direction of a vortex at the Equator? Naively one might think that there will be no vortex at all and that the water will go straight down the drain. But in a real case there is always some rotation, which develops into a nice vortex, just as it does in a bathtub at any latitude. There are amusing stories about how people who live at the Equator have found a way to earn some money. A bucket with a hole and a plug in the bottom is filled with water. When the plug is pulled, an unnoticeable manipulation by the performer forces the vortex into the desired rotation. The plug is put back, and the bucket is rushed to the other side of a line on the street, which has been painted to mark the Equator. There the plug is pulled again, and the amazed spectators now see the vortex go in the opposite direction. Then it only remains to collect the money for the show.

6.2 You See What You Want to See

The seventh wave on the sea is not bigger, Galileo Galilei made a blunder when he argued that the Earth rotates around the Sun, and a navy was hunting mink instead of submarines.

Waves Are Rolling In

Standing at the shore you watch the waves coming in, and start counting. One, two, three, four, five, six, seven. Ah, there it is again! The seventh wave is bigger. At least, this is what you may think, but it is merely another legend. Why just seven? Probably because it is a magic number in many cultures. But there are also people who claim that the eighth wave will be the biggest. If someone holds the view that every second or third wave is bigger, it is easily seen to be incorrect. The idea that every 17th or 23rd wave is bigger is not likely to catch people's attention. Gauging such a claim would require quite a long attention span, and it would not be easy to remember the height of the initial big reference wave in a pattern of waves with varying heights.

In an attempt to see if the seventh-wave legend is correct, we may stand at the shore, looking for a particularly big wave. When it reaches the shore, we start counting. As the sixth and seventh waves come in, we may note that the sixth wave was definitely bigger than the seventh one. Well, the rule of the seventh wave cannot be that precise. After all, six is as close to seven as we can come. Then we start counting again. But now the eighth wave may appear to be bigger than the seventh. Eight is also close to seven, thus giving some evidence for the claim. In this way we may talk ourselves into the popular belief. And it is not an obviously absurd idea that the seventh wave is bigger.

There are many remarkable periodic returns in nature. For instance, sunspot activity varies with a period of about 11 years. The weather phenomenon El Niño is repeated with a period of typically four to seven years. In eastern North America there are two types of

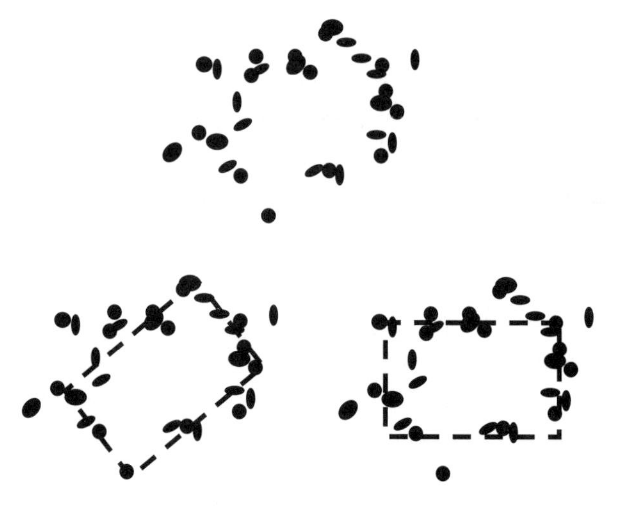

Fig. 6.3. Arrangements of stones in a hypothetical archaeological excavation (top figure) and two suggested orientations of a rectangular house (lower figures)

cicadas that emerge in large numbers every 13th or 17th year. (The latter are sometimes called 17-year locusts, although they are not locusts but belong to the order Orthoptera.) But in the case of waves on the sea, there is neither statistical evidence nor theoretical arguments that the seventh wave will be bigger.

The human brain seems to be constantly looking for simple structures. Pattern recognition has a value in the evolutionary history of struggles to survive. However, that may also lead us to "see" structures even when they are absent. Such structures can be abstract, like in an imagined conspiracy. They may also be geometrical patterns, as in the following hypothetical example. Suppose an archaeological excavation has revealed a pattern of stones arranged as in the top part of figure 6.3. They are thought to come from the foundation of a rectangular building. But how was the building oriented? The lower part of figure 6.3 shows two possibilities. Is one of them more convincing than the other? A person who has a strong belief that the layout should be as in the lower right part may claim that that there is good support for such an interpretation. Cover the lower left figure and decide for yourself. Then cover the lower right figure. Does the one on the left seem a better interpretation? Depending on what one wants too see, that particular interpretation may seem confirmed—particularly if no alternative is considered. Someone without bias may think that there is no support at all for a simple rectangular foundation in figure 6.3.

Galileo Galilei's Trial

Eppur si muove ("and still she moves") is what Galileo Galilei mumbled when forced by the Church to renounce his view that the Earth circles around the Sun. At least this is what a well-known, but most likely incorrect, legend says about Galileo's confessions at the trial. What is much less known is that although Galileo's conclusion about the motion of the Earth was correct, his argument was false.[7] Galileo wrote four books. He considered the last one, about the tides, to be his most important work, containing definite evidence for the idea about the Sun and the planets. Briefly the argument was as follows.

Galileo was familiar with the phenomenon of tides—for instance, in Venice, which he had frequently visited. There are two tides in 24 hours. During the same time, the Earth makes one revolution around its own axis. Figure 6.4 shows, schematically, the motion of the Earth around the Sun, and its own rotation. The velocity of a point on the surface of the Earth facing the Sun is the superposition of two parts—one of them in the direction of the Earth's rotation around the Sun, and one in the opposite direction due to the rotation of the Earth on its axis. At a point on the opposite side of the Earth, these two velocities are instead in the same direction. Therefore, Galileo argued, the velocity of points on the Earth's surface has a superimposed oscillation with a periodicity determined by the rotation of the Earth. That will give rise to high and low tides twice a day.

We now know that this explanation of the tides is completely wrong. Tides arise from the gravitational effects caused by the Moon, as clarified by Isaac Newton 44 years after Galileo's death. It may

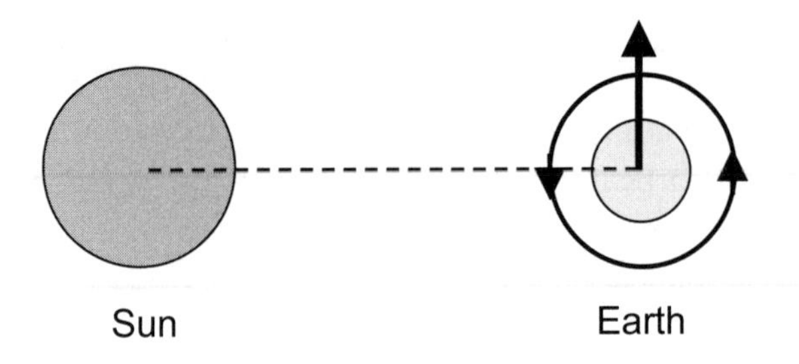

Fig. 6.4. Schematic representation of Galileo Galilei's argument for how tides arise

seem easy to understand that there is a high tide at a location which faces the Moon, but to account for the simultaneous high tide on the opposite side of the Earth, we must consider not only the Moon's gravitational pull but also the centrifugal effect as the Earth and the Moon rotate around their common center of mass, like a swirling couple in a dance. The high tide often occurs with a considerable time delay (several hours) after a location has been directly facing the Moon, but that depends on how water passes through geographical features like narrow straits, and is a technicality of no interest in our present discussion.

A small but important characteristic phenomenon is that the time between two consecutive high tides is not exactly 12 hours but 12 hours and 24 minutes. How can it be that Galileo did not recognize this as incompatible with his own explanation of the tides? It is a universal phenomenon and must have been noticeable to Galileo when he was in Venice. We can only speculate on this. Perhaps he was so convinced by his own argument that any evidence against it was unconsciously suppressed.

Submarines and Mink

On the evening of October 27, 1981, the Soviet submarine U137, with nuclear weapons onboard, hit a rock and got stuck above the water in the Swedish archipelago, close to a naval base. Thirteen years later, on May 25, 1994, the Swedish prime minister Carl Bildt wrote a letter to Russia's president Boris Yeltsin, expressing Swedish concerns about continued Russian violations of Swedish waters. However, the recorded sound that Carl Bildt referred to in his letter came from swimming mink, not submarine propellers. That story provides an excellent example of how hard scientific facts can be mixed with incorrect speculations, to give a picture that makes it difficult for the public to judge what has actually happened.

The U137 event was a wakeup call for the Swedes. Now there was proof of foreign powers operating in Swedish territory. The number of alleged observations rose dramatically. People often contacted the evening press, which gave vivid descriptions of observations of submarines. In 1987 there were no fewer than 956 incident reports registered by the Armed Forces. They were categorized on a scale from 1 to 6, where 1 is confirmed underwater activity, 5 means that there was no underwater activity, and 6 means that the evidence is insufficient for a conclusion. Most of the reports were assigned level 4 — underwater activity cannot be excluded. In practice it meant that people saw what they wanted to see.

After the fall of the Soviet Union there were only a few level 1 incidents. Then, in 1992 and 1993, the Swedish Navy introduced a new passive sonar system with microphones attached to buoys. On several occasions they recorded a sound that was interpreted as coming from a propeller. Since only one of several deployed buoys picked up that sound, it was concluded that the source must be close to that particular buoy. The water surface was constantly surveyed from a naval ship at a distance. Since no vessels were seen close to the buoy, the only reasonable interpretation was the presence of a submarine.

This was after the cold war, so there were rather good direct contacts between Sweden and Russia. The recording from the buoys was investigated by Russian experts. They concluded that it was a sound from a man-made object, most likely coming from a propeller. That statement from the Russian side, together with the Swedish Navy's knowledge about the circumstances when the recordings were made, was sufficient for the supreme commander of the Swedish Armed Forces to state in his 1994 annual report to the Swedish government that there had been several underwater violations of Swedish waters.

The National Defense Research Establishment in Sweden had a highly qualified group of scientists analyzing sound recordings. Initially they had come to the same conclusion as their Russian colleagues, but in early spring 1994 a new analysis of the buoy signals showed that the sound might have come from a source so close to the surface that a submarine propeller could be excluded. Instead, the sound was thought to have some kind of biological origin. These suspicions grew stronger during April and May 1994, but the scientists' contacts with the Armed Forces did not reach up to the highest level. Prime Minister Carl Bildt, unaware of this development, then wrote his May 25 letter to Boris Yeltsin.

At the same time, the scientists — of course, not knowing about the letter — continued their work. In early June they thought that the sound came from deer, and experiments were prepared. Then in July, and before any experiments had been carried out, personnel on Swedish Navy vessels observed minks, which were correctly associated with the sound recordings. The Swedish public got upset and thought that all the claims by the Navy that foreign submarines were operating in Swedish waters was just a ruse to get more money. The general elections in Sweden in September 1994 led to a new government. Then Bildt's letter to Yeltsin became known, and the public got even more suspicious about previous reports from the Navy.

The new government decided to appoint a commission of independent scientists to assess all the evidence for foreign underwater activities in Swedish waters between 1981 and 1994.[8] Their report, in December 1995, showed that there was undisputable proof of many such activities. Several of these activities involved sabotage of underwater installations, which must have been carried out by foreign forces. Yet, with the exception of the stranded U137, there was insufficient scientific evidence to identify the country responsible for the violations.

This story gives rise to many thoughts. It would have been remarkable if the Swedish Navy had ignored the conclusion of the Russian experts that a man-made object had caused the sound picked up by the sonar. It is thus understandable that Bildt wrote his letter of protest to Yeltsin. The scientists at the Swedish National Defense Research Establishment worked as scientists should, constantly scrutinizing the evidence before drawing conclusions. The developments in the spring of 1994, with the Swedish prime minister and the staff at the National Defense Research Establishment unaware of the other's actions, have many parallels in other events where a growing concern among specialists has not reached the highest administrative level in time (cf. sec. 2.3, The Loss of a Spacecraft). Finally, the government's appointment of a politically independent commission of respected scientists meant that the confidence of the public was restored. What had been one of Sweden's most important foreign policy issues after World War II had been put to rest.

6.3 Suddenly Something Happens

When a torn fishing net is no longer a net, why trees can't reach to the heavens, and what the physics of freezing water has in common with sociology.

Fishing Nets, Coffee Percolators, and the Web

Consider a new fishing net with threads that form perfectly square openings. With wear and tear some threads are broken. How many threads can be broken before it is no longer a net? That depends, of course, on what we mean by a net. Let us consider a large square net hanging like a hammock from bars attached at two opposite sides. The threads forming the sides of a single mesh are cut, one by one and in a random way (fig. 6.5). The net becomes increasingly useless for fishing, but it is at least connected. Then, finally, by the cut of a single crucial thread the net is divided into two parts, hanging as two smaller and severely damaged separate nets.

We can make an analogy between our fishing net and an electrical

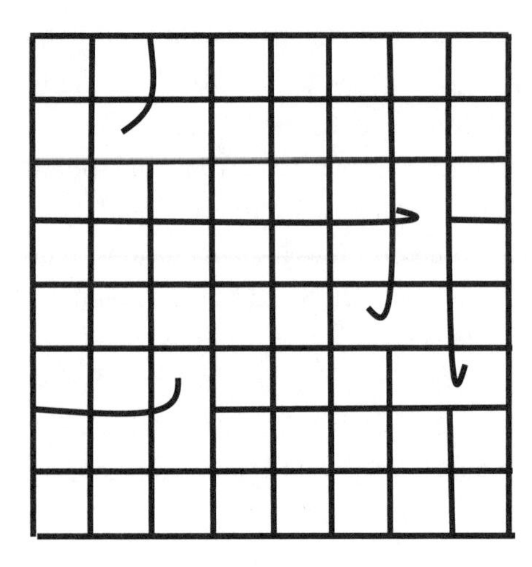

Fig. 6.5. Threads randomly cut in a net

network. Let each side of the small square openings in the net be made of electrically conducting wires, so that there is a certain electrical resistance between two opposite sides of the big net. As the conducting wires are cut, one by one and randomly, the electrical resistance between the opposite sides increases. Initially, current flows only in the wires that are parallel to the overall current direction. Cutting the orthogonal wires does not change the total resistance, but when many wires have been cut, the electrical current flows in contorted paths, and the parallel wires can also play a crucial role.

Eventually there will be a single crucial wire, like a bottleneck, through which all the current has to pass. When it is finally cut in the random cutting procedure, the net becomes disconnected and the electrical resistance between the sides of the net is infinite. In a large net this happens when a statistically well-defined fraction of the mesh sides has been broken.[9] Physicists call that the percolation limit or percolation threshold, making an analogy to how liquid finds its way through a coffee percolator. Close to the percolation limit the overall resistance varies rapidly when wires carrying a large part of the total current are cut.

As an example of the concept of percolation think of the network of highways connecting California with Massachusetts. The number of possible driving routes between the two states is enormous if we are not interested in the fastest or shortest routes. But if more and more of the roads were closed, in a random way, we would eventually find that California and Massachusetts were separated with regard to road traffic.

The modeling of percolation phenomena as networks has many practical applications—for instance, in geosciences. Deposits of oil and gas may occur in porous material in which the pores are either separated or connected through channels. The extraction of oil and gas from all these pores requires that they be linked together. Most people, however, associate networks with Internet. In 1958, the Advanced Research Projects Agency (ARPA) was established in the United States as a response to Sputnik and other technical achievements in the Soviet Union. ARPA developed a highly robust communication network, the ARPANET, with a first connection in 1969 between Los Angeles and Menlo Park in California. In 1988 the network became open to commercial activities. CERN, the European physics organization for particle research located in Geneva, released the World Wide Web. What then happened, we all know.

Goethe and the Height of Trees

The book *Wahrheit und Dichtung* (*Truth and Poetry*) by the German poet Johann Wolfgang von Goethe (1749–1832) has the motto *Es ist dafür gesorgt dass die Bäume nicht in den Himmel wachsen* ("Care is taken that trees do not grow into the sky"). There are many factors limiting the height of trees. One of them is that a high tree may buckle under its own weight.

Let us model a tree in the simplest possible way. Its trunk is represented by a long vertical column, firmly fixed in the ground (fig. 6.6). Let a force bend the column by pulling horizontally at the top. If the force is taken away, the column returns to the vertical position. The bent column (tree trunk) is in a state of higher energy. In this case, elastic energy is stored, and then released as the column straightens again. Further, the top of the bent column is slightly

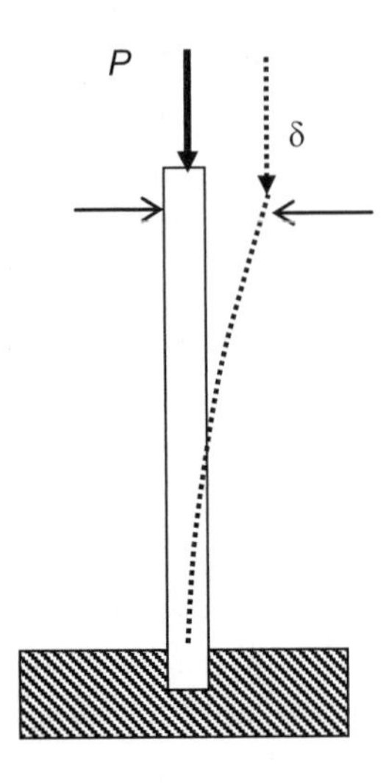

Fig. 6.6. The bending of a tree, schematically represented by a column

lowered toward the ground. That will be another key point in understanding spontaneous buckling. We now add the effect of a large tree crown in the form of a vertical force applied at the top of the column. As the column bends, this load is lowered — in other words, the potential energy of the tree crown decreases. Thus, there are two energy terms: one increasing the elastic energy in the bent trunk and one decreasing the potential energy of the crown. For light crowns, we can ignore the latter term. But with a very heavy crown, the lowering of the potential energy can be so great that the trunk bends more and more, until it breaks.

We would have to delve too far into elasticity theory to derive the pertinent formula for these events. Therefore, we will use a graph to illustrate the idea behind what is technically termed Euler buckling.[10] When the free end of the trunk is pulled horizontally a certain distance, the elastic energy increases as in the top curve in figure 6.7. Then we add the potential-energy change caused by the lowering of the load (i.e., the weight of the tree crown). If the load is large (bottom curve in fig. 6.7), the sum of elastic and potential energy becomes more and more negative as the horizontal displacement of the top of tree increases. The tree spontaneously bends and breaks. With a smaller load (middle curve in fig. 6.7), the sum of elastic and

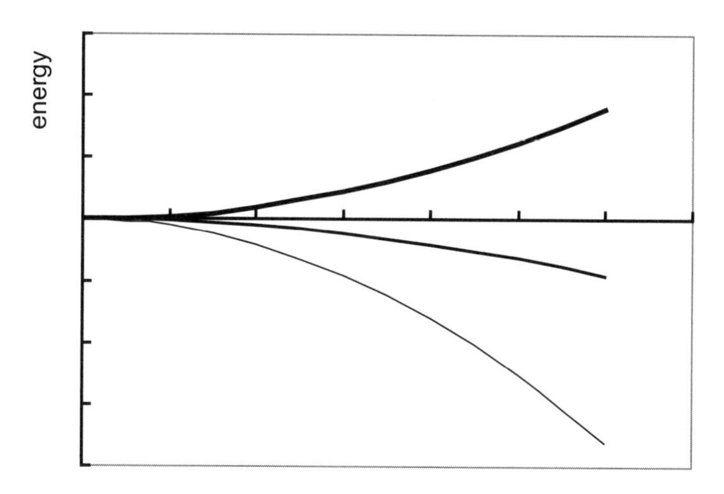

Fig. 6.7. Energy curves for a bent tree. The thick line gives the increase in elastic energy. The two thinner lines give the decrease in potential energy when the center of gravity is lowered.

potential energy would instead increase with the assumed horizontal displacement. Since nature does not allow total energy to increase spontaneously, such a bending will not take place, and the tree remains in an upright, stable position.

One might object that our model is a rather poor representation of a tree. A tree trunk usually tapers off. Instead of a crown resting on the very top of the trunk, there are branches. Further, we ignored the mass of the trunk itself. These are valid criticisms if the goal is to find, for instance, the maximum height of a particular species of tree. But a more detailed description will not alter the main conclusion — that a tree will buckle under its own weight if it is too high and slender. In nature, trees often reach about 25 % of the height that would lead to buckling under their own weight, which means a reasonable factor of safety. Bamboo, which does not have a heavy crown, is nevertheless close to the buckling limit, but bamboo is a kind of grass, not a tree.

Supercooled Rain and Critical Mass

In countries that have snow and ice in the winter, automobile drivers may experience a very unpleasant precipitation phenomenon—supercooled rain. We all know that water freezes to ice below 0 °C (32 °F). If it is colder, it will come down as snow, sleet, or hail. But sometimes, even though raindrops have a temperature below the freezing point, they remain liquid. When they hit an object like the surface of the road, they instantaneously freeze to ice. To drivers, the road may seem dry and safe. Then, suddenly, it turns into something more like the surface of a curling rink.

The physics behind the phenomenon is that of nucleation and growth. Analogous examples can be found far from the world of physics—in sociology, for instance. A population can have a silent opinion that something should be changed. Nobody wants to take the first step, perhaps because there is some kind of personal cost associated with a change or a fear of negative reactions from others. A few brave individuals may try to take steps toward a change, but lacking enough support from other people, they give up, at least temporarily. Then, seemingly by chance, a group of dissidents becomes strong enough to remain active for a while. More and more people join them, perhaps ending in a complete change among the population. It all started on a small scale, with a "nucleus" that became so large that it continued to grow irreversibly. The issue could be political, but similar changes also occur in, for instance, the world of fashion.

Returning to supercooled water and the formation of ice, we can now make a simple model. Assume that a certain mass of water has been cooled to a temperature below 0 °C. Nature seeks that state which represents the lowest energy.[11] In our case it means that the water transforms from a liquid to a solid—ice. Suppose that we have a little sphere of ice inside the volume of water (fig. 6.8). That lowers the energy. However, at the same time we have created a surface (an interface) between the new ice sphere and the surrounding water. That surface can be assigned a surface energy. Now there are two

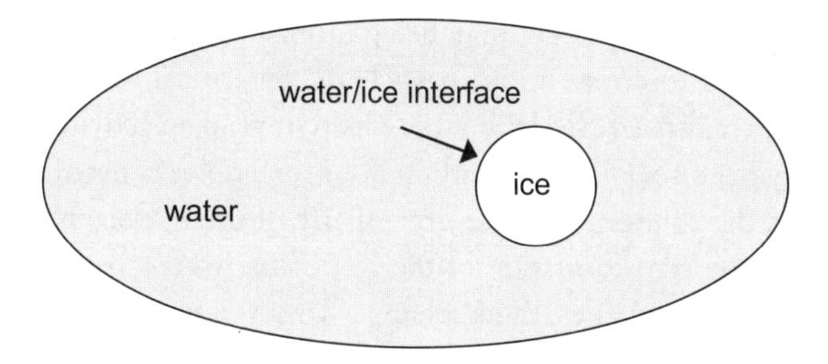

Fig. 6.8. Nucleation of a small particle of ice in a volume of supercooled water

opposing contributions to the energy — one that lowers energy due to the transformation from liquid water to ice, and one that increases energy due to the creation of a surface. For very small ice spheres, the increase in surface energy outweighs the decrease due to solidification. This is another example of the square-cube law. Imagine that we make the radius of the sphere smaller by a factor of 10. The area of the sphere, and therefore also the total surface energy, is smaller by a factor of 1/100, but the volume of the sphere, and hence the lowering of the energy due to ice formation, is smaller by a factor of 1/1000.

The molecules in supercooled water are in constant motion and "want" to arrange themselves in the configuration of an ice crystal, but the surface energy forces them back to the state of the surrounding water. Sooner or later, however, a little "nucleus" of ice by chance gets so large that it reaches the critical size. Then it may either shrink or grow; in both cases the energy is lowered. It is like being at the top of a mountain pass, with downhill paths either way. The critical size of the nucleus is the tipping point. If that has been passed, there will be rapid growth. Hence the characterization of freezing water as a nucleation-and-growth process.

This picture tells us the principle behind ice formation, but it has not yet explained why supercooled rain does not freeze until it hits an object, such as the road. The real mechanism is complex, with surface energies playing a key role. It turns out to be easier to get a nucleus of ice when it can be formed at an interface between water and another object, rather than being formed only in water. A little "seed" helps to create an ice particle of the critical size, just like inventors may need "seed money" to help start up an activity. In our analogy from sociology, we can think of something happening that initiates the change. Once the critical size of the nucleus has been reached, the transformation of the rest of the water to ice is very rapid—so rapid that we think it happens instantaneously.

The closely related concept of a critical *mass* (rather than critical size) first became known to a wider public in connection with the atomic bomb. Atoms of uranium spontaneously emit neutrons, which may hit other uranium atoms and trigger them to emit more neutrons, in an escalating chain reaction. At the same time energy is released. There must be many uranium atoms in a small region, forming a critical mass; otherwise, the neutrons may escape to the surroundings before they get a chance to hit another atom. In the atomic bombs used in World War II, subcritical pieces of uranium or plutonium were kept separated and then brought in contact to get a supercritical mass at the moment of explosion. Today, the concept of critical mass is perhaps most often used in the description of creative workplaces or the restructuring of activities. If enough people get together and exchange ideas, this may generate even more new ideas and eventually a breakthrough of some kind. When several organizations or companies merge, there can be scale effects, which lead to further growth at a lower relative cost.

6.4 Engineering versus Science

Why schoolbook descriptions of Archimedes' principle, a beam on supports, and a rope on a pulley give wrong predictions in the real world.

Slapstick

In the novel *Slapstick*, the American writer Kurt Vonnegut (1922–2007) tells us how New York is hit by sudden jolts of high gravity forces:

> The force of gravity had increased tremendously. There was a great crash in the church. The steeple had dropped its bell. Then it went right through the porch, and was slammed to the earth beneath it.
>
> In other parts of the world, of course, elevator cables were snapping, airplanes were crashing, ships were sinking, motor vehicles were breaking their axles, bridges were collapsing, and so on and on.[12]

This passage has captured the interest of physicists. Of course gravity could not vary suddenly, but ignoring that, let us momentarily assume that gravity actually did change as described. Is there anything in Vonnegut's description of the consequences that stands out as counter to physics?

Let us start with the observation that elevator cables break. Such a cable is designed to carry the elevator with its full load and with a "safety factor" that can be defined as the ratio

$$\frac{\text{actual strength}}{\text{required strength for allowed load}}.$$

The required safety factor in building codes for elevators is not a universal number but varies with, for example, the design speed of the elevator. Typically it lies between 7 and 12 (see table 3.6). If cables were to break as Vonnegut writes, the increase in the force of gravity would have to be at least a factor of 10. This is a reasonable lower limit because few elevators would be loaded to their allowed capacity when the jolt of increased gravity occurred, and Vonnegut's text suggests that breaking cables was common. The increased load on the elevator cables during acceleration is certainly much smaller than the acceleration of gravity, g, and therefore negligible compared to the change in gravity force that Vonnegut mentions.

Next we consider the case of the sinking of large ships. Archimedes' principle says that the force of buoyancy is equal to the weight of the displaced liquid. Therefore, as the weight of the ship increases in proportion to the force of gravity, so does the buoyancy, and the ship would not sink deeper. This is the physics schoolbook result. But in the real world of engineering, a large ship would probably sink.

Under normal gravity conditions the buoyancy of the ship comes from the pressure of the water. That force is distributed over the ship's entire hull. The weight of the ship, on the other hand, is the sum of the gravity forces acting on all the masses making up the ship. They are very unevenly distributed. Therefore, just as church bells would fall down and axles break on land during a sudden increase of gravity, we could expect similar breaking and distortion in a large ship. The damages to the ship would probably be so severe that the ship would sink. As a contrast we can think of a dinghy or a toy ship. These are so small that they would withstand an increase in gravity without internal breaking or distortion and would float at the same level as they did before the gravity change.

Understanding the schoolbook version of Archimedes' principle is important. But as we have seen in this example, engineering reality may have surprises for the unaware physicist who has focused only on general principles. In particular, a scaling in size (up or down) may lead into a regime of new and important physical phenomena. For instance, Archimedes' principle is of little importance for an insect on a water surface, where surface tension dominates over buoyancy.

Not a Schoolbook Problem

A horizontal beam rests on two supports as in the left-hand drawing in figure 6.9. What is the force exerted by the beam on each of the supports? This is a typical problem in introductory physics books. But let the beam rest on *three* supports. That is beyond what even advanced physics books cover. Going from two to three supports means leaving the idealized world of physics and entering the reality of engineering. No beam is absolutely rigid. No supports can be absolutely stiff, and placed at exactly the same horizontal level. When

Fig. 6.9. The figure on the left represents a physics problem and the one on the right an engineering problem.

the beam rests on only two supports, none of these complications are important. The forces on each support are always equal to half the weight of the beam, $mg/2$.

In both science and engineering, idealization is often a key to a deeper understanding. To get a first idea about the nature of the problem with three supports, we start with a uniform and absolutely rigid beam, resting on equally spaced supports at equal level. Let the supports be very compressible and represented by springs. Since the beam does not tilt to either side but remains horizontal, all three springs are equally compressed and therefore carry the same load:

$$F = mg/3.$$

Next, reverse matters and let the beam be very flexible and the supports absolutely rigid. An extreme case would be to replace the beam by a long chain. It will hang in a shape reminiscent of the inverted logo of a well-known fast food restaurant, provided of course that one prevents the ends of the chain from sliding off the outer supports. The two halves of the chain may be considered separately. For symmetry, each end of the imagined half chain is held by a vertical force of $mg/4$. The total force, F, on the middle support is $mg/2$, with $mg/4$ on each of the outer supports.

Our discussion suggests that F for a real case lies between the results for a perfectly rigid and a perfectly flexible beam:

$$mg/3 < F < mg/2.$$

However, this simple conclusion is not true. The statement that the beam rests horizontally on three supports at equal levels is an idealization in which the distances δ and ϵ in the exaggerated figure 6.10 are exactly zero. The force, F, from the middle support on a perfectly

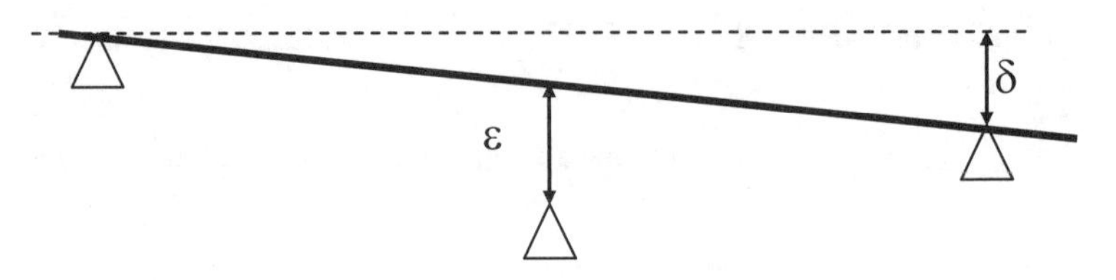

Fig. 6.10. An exaggerated figure showing that the three supports for the beam do not lie in exactly the same horizontal plane

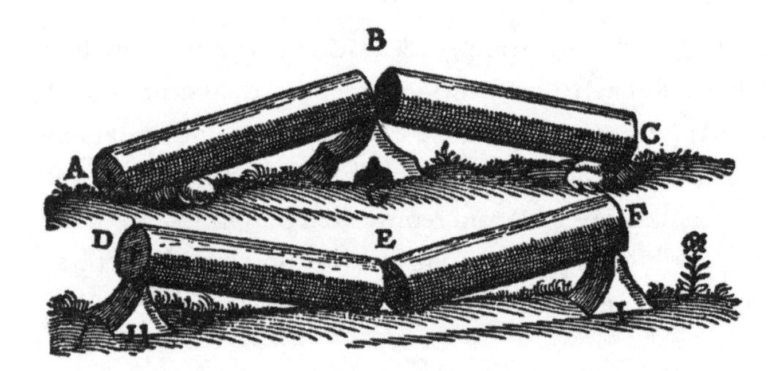

Fig. 6.11. A stone pillar breaks easily under tension. From Galileo Galilei, *Dialogues concerning Two New Sciences* (1638)

flexible beam (a chain) depends negligibly on the precise values of δ and ϵ. A perfectly rigid beam would have $F = 0$ if $\epsilon > 0$ as in the figure, and $F = mg$ if $\epsilon < 0$ so that the beam balances on the middle support. Thus, we can only conclude the trivial fact that the middle support carries a load, F, with

$$0 < F < mg.$$

A real beam is neither perfectly rigid nor perfectly flexible. We need a measure of the deflection (sagging) under its own weight. If the beam is so flexible that it sags more under its own weight than the small deviations of the supports from a perfect horizontal level, elasticity theory gives that the force on each of the outer supports is $3mg/16$. Hence, the force on the middle support is

$$F = 5mg/8.$$

Our discussion shows that a system as simple as that of a beam resting on three supports is far from easy to analyze. Depending on which extreme case we consider, the force on the middle support can be 0, $mg/3$, $mg/2$, $5mg/8$, or mg. Books on elasticity theory usually give the result $5mg/8$ because that corresponds to the most common engineering situation.

We end by noting that as early as the seventeenth century Galileo Galilei, in his book *Dialogues concerning Two New Sciences* (1638), discussed the difference between different kinds of support under a horizontal stone pillar. Stone can withstand large compression forces but breaks easily under tension (fig. 6.11).

Hoisting a Sack

When muscles, and not machines, were the prime movers in everyday life, it was important to use simple principles of mechanics to make work easier. The lever may be the best illustration. Figure 6.12 gives another example—the pulley. It shows the application of a fundamental principle that is still taught in elementary physics courses.

But many of today's youngsters have no personal experience of hoisting a load with the help of ropes and pulleys, so a teacher might instead ask what the required force will be if the rope in figure 6.12 does not go through a pulley but instead lies over a smooth horizontal branch of a tree. The mass of the load is given as 25 kg. That teacher may be unaware that no ordinary young girl or boy can hoist such a load. This is because even if the branch is said to be smooth, friction cannot be ignored.

There is a very simple formula that gives the forces on a rope in contact with a cylinder. In spite of its elementary mathematical form, it is usually not included in science courses, and surprisingly, is also unknown to many engineers. With notation shown in figure 6.13, it reads

$$F_2 = e^{f\varphi} F_1,$$

Fig. 6.12. A common illustration in physics books a century ago

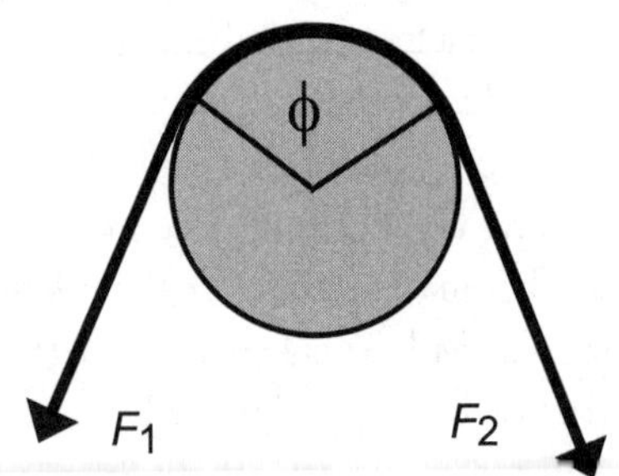

Fig. 6.13. Geometry of a rope over a circular beam with friction

where $e \approx 2.7$ is the base of natural logarithms, ϕ is the angle in radians that the rope makes along the circumference of the cylinder, and f is the friction number for the rope against the material of the cylinder. F_2 is the minimum force required to pull the rope if the force at the other end is F_1.

Although the branch was characterized as smooth, the friction number for rope against wood will be at least in the range of 0.3 to 0.4. One full turn around a cylinder subtends the angle 2π radians. In our case ϕ is half a turn, or about 3 radians, so $f\phi$ is about 1. Thus, to hoist a load corresponding to F_1, we typically need a force at least as large as

$$F_2 \approx 3F_1.$$

If the mass of the load is 25 kg (55 lb), it can be raised only if the mass of the girl or boy pulling the other end of the rope is at least about 75 kg (165 lb), and if she or he is then dangling from the rope to maximize the downward force F_2. The common schoolbook instruction "neglect friction" seldom has any relevance in real life.

7
Tricks of the Trade

7.1 A Crash Course in Science Thinking

How scientists rely on thought experiments to shed light on problems about a dinghy in a pool, a faulty escalator, and a floating apple.

Dinghy, Anchor, and Pool

One of the most popular physics brain twisters goes something like this:

> *A dinghy, with a passenger and a heavy anchor, floats in a small pool. The passenger throws the anchor into the water. Will the water level in the pool increase, decrease, or stay the same?*

This problem is an ideal one to illustrate how we can get insight into a problem by considering special or extreme cases. The assumptions need not be realistic as long as physical laws are not violated. We will analyze what is called a *thought experiment* or, in German, a *Gedankenexperiment.*

Let the mass of the dinghy plus the passenger be negligibly small. Further, let the anchor be as small as a lead shot, made not of lead but of a material with an extremely high density. In fact, we assume the mass of the anchor to be so large that the dinghy barely floats. When the dinghy is placed in the pool, a volume of water equal to that of the dinghy is displaced, thus raising the water level in the pool.

Next, the heavy but miniscule anchor is thrown overboard. Because it is so small, we can ignore its effect on the water level when it lies on the pool bottom. The dinghy with its passenger, assumed to have negligible mass, now floats high without displacing any water. Therefore, less water is displaced when the anchor is thrown overboard. As a consequence, the water level in the pool is lowered, which is also the answer one is supposed to give.

Now that we have found the conventional solution to the brain twister, we may ask if it has any physical significance or if it is just one of those esoteric problems that physics teachers love to give to their students. How much would the water level increase if we had a real pool, dinghy, and anchor?

When one is dealing with water, it is easiest to use SI units in estimations, since the volume of 1 kg of water is 1 dm^3. Suppose that the anchor's mass is 15 kg. According to Archimedes' principle, the buoyancy force on a floating object is equal to the weight of the water that is displaced by the object. Before the anchor is thrown overboard, the dinghy therefore displaces an extra amount of water with a mass of 15 kg and hence a volume of 15 dm^3. If the anchor is made of iron, it has a volume of about 2 dm^3. This is also the volume displaced by the anchor when it rests on the bottom of the pool. The difference, $(15 - 2)\ dm^3 = 13\ dm^3$, causes a lowering of the water level. Even in a very small pool with an area of 13 m^2 (the size of a bedroom), the level would change by only 1 mm (1/24 inch). This is about the same as the decrease in the water level due to the thermal contraction of the water if the temperature drops from 25 °C to 22 °C (from about 77 °F to 72 °F) and the depth of the pool is 1.5 m (5 ft).

Obviously, the dinghy problem is unrealistic in the sense that the effect is too small to be observed in a real situation. Is the thought experiment also so unrealistic that it leads to false conclusions? We started with some extreme assumptions about densities. Considering that the densities of real materials do not exceed 23 000 kg/m^3 (the density of the element osmium)—23 times that of water—one may wonder if it is good physics thinking to assume an almost infinite density. The answer is yes: we are free to imagine any density. There is no compelling law of nature that puts a limit on densities in our case. Sometimes it may require a good knowledge of the physical world to decide if an assumption in a thought experiment is acceptable. For instance, one can imagine a slab that shields from magnetic forces,[1] but there is no material that can shield from gravitational forces.

Up and Down the Escalator

A subway station has escalators going up and down. One day the escalator that goes down is out of order. That provides an opportunity for an experiment while you are waiting for the train to arrive. You ask your friend to take the ascending escalator to the top and then down again, walking on the escalator the entire time. Simultaneously you will walk up and then back down the escalator that is idle. Who will be the first to return to the platform if both of you walk with the same speed relative to the escalator steps?

Again, it is illuminating to consider an extreme case. Suppose that your friend walks at a speed that is *lower* than that of the escalator itself. Going up will be faster than on an idle escalator, but on the return she is carried backward faster than she can walk and remains "stuck" at the upper end. With a walking speed just a little bit faster than that of the escalator, she is able to return to the starting point, but it will take a long time. This argument gives the trend. It seems obvious that it is always slower to take the moving escalator up and down. A student might find this to be sufficient as an answer to the problem of who will be the first to arrive back, but an engineer should also ask if it matters in practice.

Now let us try some realistic values. Assume that the length of the escalator is 30 m. Typical escalator speeds are between 0.5 and 1 m/s, and we take 0.5 m/s. The walking speed relative to the escalator itself is assumed to be 1 m/s, both up and down. If the escalator is not moving, it will take 30 s to walk in each direction, or a total of 60 s. For the ascending escalator we add the speed of walking (1 m/s) and the speed of the escalator (0.5 m/s). To cover 30 m then requires 30/1.5 s = 20 s. On the return, the effective speed is 0.5 m/s and it takes 60 s to cover 30 m. The total required time, up and down, is 80 s. That is 20 s longer than with the idle escalator — a noticeable difference.

After these preliminaries we have a good grip on the problem and are ready for a mathematical solution. Let the walking speed be u relative to the surface you step on (the moving or the idle escalator steps), and let the speed of the escalator relative to the surroundings be v. The distance between the end points of the escalator is L. The speed relative to the surroundings is $u + v$ when walking on the escalator in its direction of motion, and $u - v$ in the opposite direction. The total time needed is

$$t = \frac{L}{u+v} + \frac{L}{u-v} = \frac{2Lu}{u^2 - v^2} = \frac{2L}{u} \cdot \frac{1}{1 - v^2/u^2}.$$

The last factor is always larger than 1 (if $u > v$), so t is always larger than the time $2L/u$ it takes if the escalator is idle. An interesting case is the one in which $v > u$. Then the time, t, becomes negative. One might object that this is unphysical, but instead of just discarding the solution on this ground, we can provide an interpretation for it. When $v > u$, one either gets "stuck" on the escalator or can't "take off," depending on the end of the escalator at which one starts.

There are many variations on this problem. A common version considers an airplane flying from A to B and back, when there is steady headwind or tailwind. A more mundane example is that of a bicycle ride to and from a destination on a windy day. In both cases the round trip takes longer if the wind is blowing. Likewise, it is more difficult to run 10 000 m in a stadium if a steady wind is blowing than if it is calm. The extra resistance on the parts of the track where there is a headwind is not fully compensated for on the parts where there is a tailwind.

Note that in the case of the escalator we postponed finding a mathematical solution until we had a good understanding of the problem. The reason was to stress the mode of thinking used by good problem solvers. Novices tend to rush away and write equations too early. Even if the mathematical solution is formally correct, one can easily fail to recognize and understand certain features, such as the interpretation of a negative time t in our case.

The Floating Apple

Here is a slightly changed version of a problem that has been given in an entrance exam to Finnish universities of technology.

> *A beaker is two-fifths filled with water. An apple floats in the water, with a certain fraction of the apple's volume submerged in water (fig. 7.1). Then one adds as much olive oil as there is water in the beaker. How large a fraction of the apple's volume is now submerged in water? (a) A larger fraction. (b) A smaller fraction. (c) The fraction is unchanged.*

The first thing to note is that water and olive oil don't mix, and the lighter oil will float on top of the water. One might then think that alternative (a) is correct because the oil pushes down on the apple, so that it sinks deeper into the water. On the other hand, one could argue that alternative (b) is correct because the oil pushes down on the water, increasing the pressure in the water and giving a larger uplifting force on the apple. A third possibility may seem to be that the two effects just mentioned will cancel and that alternative (c) is correct. When this problem is discussed in physics classes, it sometimes leads to heated debates, with all three alternatives getting proponents. Then an ambitious student may suggest that one should apply Archimedes' principle to formulate and solve the proper equations. However, there is a simpler approach if we only want to decide which of the alternatives, (a), (b), or (c), is correct.

Look at what is happening in an extreme case. Let us replace the olive oil with water. Imagine that the original water in the beaker had been colored red, and the added water is colored green. Further, assume that the green water can be added so carefully that it floats on top of the red water. In practice, this may be impossible, but as a thought experiment it works. Since the buoyancy force has nothing to do with color, it is obvious that the apple will rise to the surface of the green water. That is alternative (b). Then we proceed in our thought experiment and add a liquid whose density is very slightly

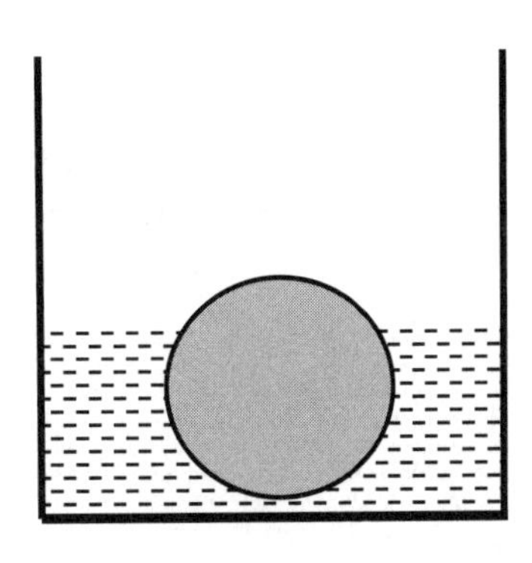

Fig. 7.1. Schematic representation of a floating apple

lower than that of the green water. The outcome will be almost the same as with green water added, and alternative (b) still holds. Finally we replace the green water with olive oil. There is no reason to think that the tendency for the apple to rise will be changed. Since the problem did not require us to calculate in detail how much of the apple was submerged in water but only to say how the submerged fraction changed, the problem has been solved.

A real professional may try different extreme cases to get further insight. The geometry of the apple is approximately spherical, but even that shape is a bit troublesome in an accurate calculation. We could therefore investigate what would happen if the apple were replaced by a slab, like a very flat box.[2] If we know the density of the slab material, it is not difficult to apply Archimedes' principle before and after oil has been added. The weight of the slab is equal to the weight of the volume of liquid displaced by the slab. Therefore, we can calculate how deep the slab floats in the water. The result is easy to obtain even if the amount of olive oil is small — yet another special case, since we originally assumed that the amount of olive oil was equal to the amount of water.

7.2 Is the Formula Accurate Enough?

On why the body mass index is unfair to strong people, the wind chill temperature is not a temperature, and ship tonnage is not a mass.

Obesity

Obesity is a major health hazard in many countries. But who should be characterized as obese? There is a need for a measure that is very simple to determine but still contains enough information to be useful. The body mass index (BMI) is often used for this purpose. It is defined as

$$\mathrm{BMI} = \frac{M}{L^2},$$

where M is the body mass expressed in kilograms and L is the height expressed in meters. Table 7.1 gives some values of BMI when body mass and height are expressed in SI units and the equivalent units used in the United States and the United Kingdom.

There are two obvious questions to ask about the BMI: What is the threshold for obesity, and should not BMI be calculated from the cube rather than the square of the height?

In Europe the thresholds are generally accepted to be a BMI > 25 kg/m^2 for overweight and a BMI > 30 kg/m^2 for obesity. In the United States several thresholds have been used, differentiating between both sex and age groups. The National Health and Nutrition Examination Survey (NHANES II) put obesity at BMI > 27.8 kg/m^2 for men and at BMI > 27.3 kg/m^2 for women. The National Academy of Sciences' diet and health report suggested that the thresholds for overweight should be 27 kg/m^2 for persons 45 to 54 years of age, 28 kg/m^2 for persons 55 to 65 years of age, and 29 kg/m^2 for persons over 65 years. In 1997 the World Health Organization adopted the European scale for both men and women in the age group 29–69. The other numbers mentioned above are no longer in use, but they

Table 7.1. Body mass index (BMI) for some combinations of mass and height

M (kg)	60	60	70	70	80	80	90
M (lb)	132	132	154	154	176	176	198
L (m)	1.60	1.65	1.70	1.75	1.75	1.80	1.80
L (ft)	5.2	5.4	5.6	5.7	5.7	5.9	5.9
BMI (kg/m^2)	23.4	22.7	24.2	22.9	26.1	24.7	27.8

illustrate the difficulty and arbitrariness in defining simple limits, be it overweight or exposure to radiation, noise, or toxic substances.

A newborn baby typically has a mass of 3 to 4 kg and a height of 0.50 m. That gives a BMI between 12 and 16 kg/m^2, far below BMI = 18.5 kg/m^2, which has been suggested as a threshold for underweight. Like so many other simple mathematical relations in science and engineering, BMI has a restricted range of applicability and should not be used outside the range for which it is constructed. To circumvent this problem health authorities decided to keep the original definition of BMI but to construct tables where the thresholds vary up to the age of 18 years.

The most common objection to BMI is that it does not take the *cube* of the height. Humans are after all three-dimensional, so the argument goes. Since there is a certain stigma in some societies about being characterized as obese, the debate can get quite fierce. In particular, it is said that the BMI is unfair to tall people. Let us therefore try a generalized BMI(p), defined as

$$\text{BMI}(p) = \frac{M}{L^p},$$

where the exponent p lies in the interval $1 \leq p \leq 3$. The whole idea behind the BMI is to be able to determine if there is a statistical correlation between BMI and certain health problems. At the same time, we want a simple definition. Thus, it is natural to look for such

Table 7.2. Density of three components of the human body

Substance	*Density* (kg/m^3)
Fat	928
Muscles	1058
Bone, compact	1900

statistical correlations when p is chosen to be 1, 2, or 3, although a fit of the body mass M to the height L for a large adult population gives a value of p between 2 and 3. When BMI(p) is correlated with health problems, it turns out that $p = 3$ is not a better indicator than $p = 2$. Therefore, we cannot refer to a simple scaling argument and claim that it is incorrect to use $p = 2$.

BMI has also been criticized as being unfair to those who are muscular, since the density of muscles is higher than the average density of the body (table 7.2). Again, we must remember that the purpose of the BMI is to get a measure of general health risks. There are few individuals who are highly muscular, and their health risks may deviate from those of the rest of the population. The variation in average body density in the ordinary population is quite small and may correspond to a change in BMI by about one unit—certainly not significant in view of the arbitrariness in the thresholds for overweight or obesity.

Wind Chill Temperature

In cold weather there is a risk of frostbite. How large the danger is depends not only on temperature but also on wind speed and air humidity. A weather reporter might say, "The actual temperature is 0 °C (32 °F), but it feels like −10 °C (14 °F)." The lower value is called the wind chill index or wind chill temperature, T_{wc}. There have been many attempts to establish a relation between T_{wc} and wind speed. In a recent model, worked out in the United States and Canada, the wind chill temperature is given as

$$T_{wc} = \{13.12 + 0.6215\, T_{air} - 11.37\, v^{0.16} + 0.3965\, T_{air}\, v^{0.16}\}$$

or

$$T_{wc} = \{35.74 + 0.6215\, T_{air} - 35.75\, v^{0.16} + 0.4275 T_{air}\, v^{0.16}\}.$$

The first formula, adopted in Canada, has T_{air} expressed in degrees Celsius and wind speed, v, in kilometers per hour, while the second formula, adopted in the United States, has T_{air} in degrees Fahrenheit and v in miles per hour. The wind speed is measured at a height of 10 meters above the ground (the usual height referred to in weather forecasts), but in fitting the formula to experiments on humans, v was multiplied by a factor of 2/3 to represent the wind chill effect at the height of the face.

The mathematical expressions above for T_{wc} are not dimensionally correct, since one cannot add, for instance, a temperature $0.6215T_{air}$ and a number 13.12, not to mention a wind speed to the power 0.16. In fact, the relations are not meant to be dimensionally consistent. Early versions of them were based on a consideration of heat losses from the human skin, expressed as watts per square meter. However, it was thought that the public would not understand such an unfamiliar unit. The formula was therefore manipulated to give an equivalent number expressed as a temperature, in degrees Celsius or Fahrenheit. The new expressions given above will also produce a number that is often quoted as an equivalent apparent temperature ("It is 25 °F and feels like 15 °F"). However, scientists stress that wind chill is not a temperature but an index that measures the human response to temperature and wind. It is the basis for wind chill charts with messages like "If the air temperature is 0 degrees Fahrenheit and it is blowing 15 miles per hour, you will get frostbite in 30 minutes."

Leaving the dimensional problem aside, there is another aspect of the formulas that can give rise to much confusion. Are they really correct as written above, or is there a misprint somewhere? The large number of digits in the numerical factors, and the "uneven" value 0.16 of the exponent, may give the impression of high accuracy. In fact, the formulas are just a mathematical fit to an underlying, partly empirical and rather crude model. The index (the wind chill "temperature") is not meant to be accurate to more than about one or two units.

Let us take the special case of a temperature at the freezing point, $T_{air \sim}$ 0 °C (32 °F). With no wind ($v = 0$) we get T_{wc} = 13 °C (55 °F). That is, of course, nonsensical. Further, the two terms containing the wind speed cancel out in the first formula above if the air temperature is 11.37/0.3965 °C = 28.7 °C. Thus, the formula seems to imply that on a rather hot day (T_{air} = 29 °C or 84 °F) it will always feel like 31 °C (88 °F), irrespective of any wind. Obviously, the formulas are relevant only for certain ranges of wind speed and air temperature. They are meant to describe the risk of frostbite in cold weather. The wind speed must be at least about 3 mile/h (5 km/h), reflecting the fact that the formulas apply to people who are moving.

There is a similar measure of the risk of fatigue, heat cramp, and heat stroke in hot weather, calculated from an expression that contains the actual air temperature and the relative humidity. The result is again an index—the heat index—and not a temperature, but it is nevertheless often communicated to the public as an "apparent temperature."

The Size of a Ship

There are few measures more confusing than those giving the size of a ship. For a long time ship sizes were expressed in register tons, where one register ton is not a mass but a volume: 100 ft^3 = 2.83 m^3. The historical reason for the name is that *tonnage* once was the name for the tax on wine casks called *tuns* (from French *tonnes*). These casks contained about 252 gallons (UK), corresponding to a mass of about 2240 pounds. The size of a vessel was then expressed by how many such casks it could carry. The measurement of register tonnage has been complex, distinguishing between gross register tonnage and net register tonnage, depending on what parts of the interior volume (like crew space and the engine compartment) are excluded.

A clear definition of the size of ships is important, since it is the basis for safety requirements, fees, harbor dues, and the like. The International Convention on Tonnage Measurement from 1969 is now used for all ships built after July 1982. The gross tonnage, GT, is calculated as

$$GT = \{0.2 + 0.02 \log_{10} V\} V,$$

where V is the numerical value of the total volume of all enclosed parts of the vessel, expressed in cubic meters. Thus, the tonnage represents a volume and cannot be given in the unit ton (1 ton = 1000 kg), which is a mass. Since there is a gross tonnage, we also expect a net tonnage. The latter is the volume of the ship that can hold cargo.

Measures of the size of sailing boats for competitions can be equally confusing. Just as a commercial vessel of tonnage 10 000 does not weigh 10 000 tons, a sailing boat in the 12-m class may be longer than 12 m, and just as the measures of commercial vessels have been revised many times, so have the rules on how to measure the size of sailing boats. Version 5.0 of the International America's Cup Class Rule used a formula to limit the size of the boats in that race. It is

$$\frac{L + 1.25\sqrt{S} - 9.8V^{1/3}}{0.686} \leq 24.000 \text{ m}.$$

Here, L is the rated length, S the rated sail area, and V the ship's displacement of water. The formula is dimensionally correct, since all the quantities — $L, \sqrt{S}$, and $V^{1/3}$ — are lengths, and SI units can be used throughout.

In 2009, the rule was changed again. This time it was not expressed as an explicit formula but was defined through a very elaborate set of rules that give some freedom in the construction of the boat but also impose strict size limits. For instance, while the old rule did not put an upper limit to the overall length and the length of the waterline, the new rule limits them to 26 m. The old rule had a limit to the spinnaker area (512 m^2), while the new rule limits the mainsail area to 225 m^2.

7.3 Characteristic Quantities

The meanings of "shallow water," "deep in the ground," and "heavy ball" all depend on what you compare them with.

How Deep Is Deep?

The waves that are rolling toward a shallow beach come in parallel to the shoreline, even if the waves further out move in a different direction. When the shoreline bends, so do the wave fronts. The physics is the same as in the refraction of a light wave that enters an optically denser medium. There, it propagates with a lower speed and is bent toward the normal to the interface. Waves on the sea also slow down in shallow water. But what is meant by shallow, more exactly?

Science and engineering are full of qualitative statements like a long bar, a large mass, a fast reaction, a low pressure, and so on. The expression "a long bar" has a meaning only if we compare with something that is also characterized by a length. If a mass is described as large, it must be put in relation to some other mass. For instance, the mass of a proton is large compared with the mass of an electron, but not compared with the mass of a tennis ball. In exam problems, the phrase "a small light source" indicates that it does not matter exactly how small the source is. Students know that one should then consider it to be point-like.

When we say that waves come in parallel to the shoreline if the water is shallow, we must compare the depth with another length. At least two quantities that characterize a wave on water immediately come to mind. Does the water become shallow when the depth is no longer large in comparison with the wavelength, or is it in comparison with the wave height? The answer is not obvious. Perhaps

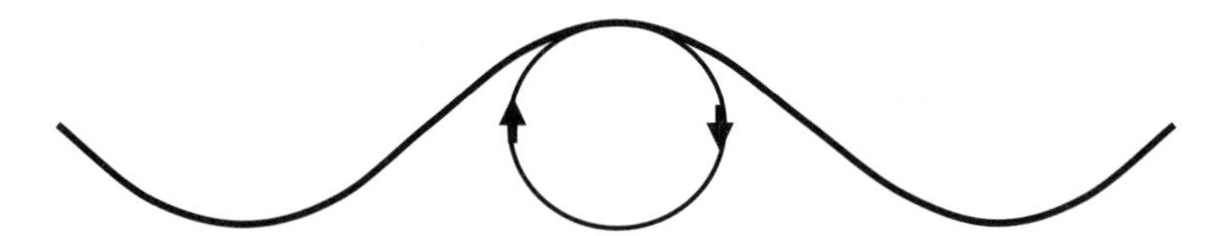

Fig. 7.2. A "parcel" of water at the water surface moves in a vertical circular path

the two quantities are closely related, so that it does not matter what we choose. It is known, however, that when the wave height is larger than about 15 % of the wave length, the wave crest will break. Therefore, we cannot have a large height without a large wavelength. But the opposite is not true. The wave height can be very small even when the wave length is large. Extreme examples are tsunami waves far from coastlines and tidal waves.

To proceed we must know more about how water moves in a wave. It may seem that water is constantly transported in the direction of the wave, but this is wrong. Instead, water moves in a vertical circle, so that the surface oscillates up and down. We may think of the well-known fact that a small floating object, like a wooden plank, does not move across the sea with the wave speed but drifts very slowly.

The top layer of water in the ocean moves as illustrated in figure 7.2. Since the diameter of that circular motion gives the wave height h, the figure suggests that h is the quantity to be compared with the water depth, but that is not correct. Water below the surface also has an oscillatory motion, but it decreases in extension with the distance to the surface. At a depth of a wavelength below the surface, there is very little left of the oscillation that we associate with the wave. A submarine can ride out a heavy storm without being tossed around by the waves. One can say that the wave "sees" to a depth of about one wavelength, and not to the depth of the wave height. Therefore, the wavelength can be taken as a crude measure of what is meant by shallow water. For the normal waves that are generated by wind, the water is shallow only close to the shore.

There are, however, waves for which even the deepest oceans are shallow—the tidal waves. Their wavelength is comparable with the radius of the Earth!

The Coldest Day of the Year

At the Greenwich observatory, east of London, there is a long time series of measured temperatures. One year, the coldest day occurred on March 27, and the warmest day on September 30. This is about two months later than one would expect for both of these temperature measurements. Surely such an extremely late winter and late summer would be something everyone had heard about—a year with a prominent place in history books. But it was in fact a normal year! These temperatures were not measured in the air at the ground level but at a depth of 4 m below the surface. There, temperature variation during a year is very small, perhaps only ±1 °C.

This is in agreement with the experience that the temperature in deep caves is almost constant and equal to the average surface temperature over the year. There is also a time delay between temperature variations at the surface of the earth and at great depths. We may compare this phenomenon with what happens when we broil a steak in the oven and then put it on the table when it is ready. The temperature in the center of the steak lags behind that at the surface, both when the steak is heated up in the oven and when it is cooling down on the table (cf. also sec. 5.3, The Age of the Earth). Here, we will concentrate on how variations in surface temperature disappear with increasing depth and defer the mathematics of the time delay to the notes.[3]

In the discussion of waves rolling in toward a beach we found that "deep water" means a depth larger than a characteristic length, which was identified with the wavelength. What is the characteristic length to use for comparison when we say that "deep below the surface of the ground, the temperature is approximately constant"? It should have something to do with how heat is conducted in the ground.

The relevant property is called thermal diffusivity, denoted a. That is a quantity that depends on the material through which heat is diffused. Thus it is different for clay, porous sand, or solid granite, but we can take a "typical," or "characteristic," value to be

$$a = 1 \times 10^{-6}\ \mathrm{m^2/s}.$$

Our problem is to identify a length that can be used as a kind of yardstick when we discuss what is meant by "deep in the ground." The thermal diffusivity itself cannot be used, since it has the SI unit $\mathrm{m^2/s}$. But if we multiply the thermal diffusivity, a, by something that is a time and then take the square root of the result, the unit becomes the square root of $\mathrm{m^2}$ — i.e., a length.

The next step is to identify such a characteristic time. Since we are interested in temperature variations caused by the changing seasons, the natural choice is the length of a year,

$$T_{\mathrm{year}} = 365 \times 24 \times 60 \times 60\ \mathrm{s} \approx 3.15 \times 10^7\ \mathrm{s}.$$

Now a and T_{year} can be combined to a give a quantity with the dimension of a length:

$$L_{\mathrm{year}} = \sqrt{aT_{\mathrm{year}}} = \sqrt{(1\times10^{-6})(3.15\times10^{7})}\ \mathrm{m} \approx 5.6\ \mathrm{m}.$$

Scientists would say that this sets a length scale for the problem. At depths that are in some sense large compared with 5.6 m, the seasonal variations in temperature are utterly small. Likewise, the variations will be significant at depths that are small compared with 5.6 m. In regions like Canada, the northern United States, and northern Europe, we may think of how the ground is frozen during the winter to a depth of a few decimeters (1 ft) or so. How we should interpret "larger than" and "smaller than" depends on what we mean by "significant" or "negligible" temperature variations. For instance, 4 m is certainly not larger than our characteristic distance 5.6 m, but in a practical case it may nevertheless be large enough that we can consider the temperature at a 4-m depth as constant.

In addition to the cyclic variation of surface temperatures during a year, there is daily variation. That defines a time T_{day} as

$$T_{day} = 24 \times 60 \times 60 \text{ s} = 8.64 \times 10^4 \text{ s}.$$

With this value we get a characteristic length:

$$L_{day} = \sqrt{aT_{day}} = \sqrt{(1\times10^{-6})(8.64\times10^4)} \text{ m} = 0.29 \text{ m}.$$

The interpretation is analogous to that for the annual variations. On a sunny day, the heat from the sun warms up the surface of a road, but the heat penetrates less than a few decimeters (1 ft) before it starts to cool again, in a cyclic diurnal variation. These rapid variations take place on a length scale, L_{day}, and a time scale, T_{day}, that are completely separated from the length scale, L_{year}, and time scale, T_{year}, of the annual variations.

Galileo Galilei, Basketball, and Table Tennis

According to legend, Galileo Galilei performed an experiment from the Leaning Tower of Pisa. When he dropped spheres of equal size but different mass from the tower, they reached the ground simultaneously. This was contrary to the ancient Aristotelian view that heavier bodies would fall faster. Galileo was not the first to challenge Aristotle; several others had come to the same conclusion in the preceding century. But it is the greatness of Galileo, and the fascination with an experiment performed from the famous building, that has made this story part of general knowledge. Many people also know that the alleged outcome of the experiment is not quite true. Heavier bodies do fall faster, as is evident if one compares the fall of a lead sphere and a feather. Thus, it was appropriate to perform such an experiment on the Moon, where there is no atmosphere. At the end of the Apollo 15 mission in 1971, Commander David Scott was televised as he took a feather in his left hand and a hammer in his right, dropping them simultaneously to the surface of the Moon. Of course, they fell equally fast.

The air drag (air resistance) on a falling object depends on the object's speed and on its cross-sectional area perpendicular to the direction of motion.[4] Compare the fall of two spheres that are made of the same material but are of different sizes, and apply the square-

Table 7.3. Data for the ball (shot) in six sports

Sport	*Mass* M (kg)	*Radius* R (cm)	*Distance* L (m)	M^* (kg)	M/M^*
Golf	0.046	2.1	150	0.3	0.2
Tennis	0.057	3.3	40	0.2	0.3
Table tennis	0.0027	2.0	4	0.006	0.5
Baseball	0.145	3.7	18	0.1	1
Basketball (men)	0.6	12	5	0.3	2
Shot put (women)	4.00	10	20	0.8	5

cube law. If the radius of the sphere is doubled, the cross-sectional area, and therefore also the retarding force, increase by a factor of 4. The weight of the sphere, however, increases by a factor of 8. Therefore, the larger and hence the heavier body falls faster.

Now let us look at the role of air drag in some sports where a sphere follows a trajectory through the air. Table 7.3 shows the radius and mass of the balls and sphere used in six sports. Is a golf ball heavy, and if so, in what sense? The adjective "heavy" has a meaning only in comparison with the mass of another object. The cross-sectional area, A, and the speed, v, are essential in an account of the retarding force from the air. However, the speed is a quantity that we may not know very well. On the other hand, we know the distance, L, covered by the sphere along its trajectory (table 7.3). Since the problem deals with motion in air, its density, ρ (expressed in kg/m^3), should also be important. The quantities A, L, and ρ can now be combined in a quantity M^* with the dimension of mass:

$$M^* = AL\rho.$$

The product AL is the volume of an imaginary tube cut out in the air by a sphere of cross section A as it moves the distance L. The quantity M^* is the mass of the air inside that tube. If one solves Newton's equation of motion for a sphere with mass M in the presence of air drag, one can show mathematically that drag is small as long as M^* is small compared with M. This identification gives a very good intuitive understanding of what is meant by a heavy or a light object in this context.

The fourth and fifth columns in table 7.3 give characteristic distances, L, and the mass $M^* = AL\rho = \pi R^2 L\rho$. Several of the measures for the balls are specified in the rules as upper and lower limits; the numbers in the table are typical values. Further, the air density depends on pressure and temperature, and the distances, L, are also typical values. In spite of these uncertainties, it is obvious that $M/M^* >> 1$ for basketball and shot putting, in agreement with our anticipation that air drag is not important in those sports.

7.4 Impress Them!

Three clever ways to use mental arithmetic and to check formulas in physics and mathematics that you have never seen before.

What Is Your BMI?

Your friend has read one of those alarming reports about health problems related to lifestyle. She says that her weight is 70 kg (154 lb), but her BMI is only 23, which is less than the threshold value of 25 for being overweight. "What is your BMI?" she asks. "We have the same weight but you are 5 cm shorter." You answer that you don't know but will find out when you get home and have access to a calculator. The BMI calculation, in which one divides by a value squared, seems to be more than one can do by mental arithmetic. But anyone with an elementary knowledge of mathematics can quickly do the calculation without calculator, or paper and pencil. It is a simple application of a rule of thumb.

The formula for body mass index is

$$\text{BMI} = \frac{M}{L^2},$$

where M is the body mass in kilograms and L is the height in meters If L increases by 1 % but M stays the same, BMI will decrease by 2 %. Similarly, if L decreases by 2 %, BMI will increase by 4 %, and so on. You double the percentage change in height to get the percentage change in BMI. This is because the BMI formula contains the square of the height.

Returning to the example, you are 5 cm (2 in) shorter than your friend. That is about 3 % in height, so your BMI is 6 % larger than 23, or about 24.5. The calculation is not exact but a very good approximation. We have just illustrated a method that is often used by engineers. For instance, they may be involved in a lively discussion about the effect of a small change in a certain model parameter. The other engineers will certainly be impressed if one of them can give a quick answer, without the help of calculators. Here are two more examples.

A cubic box can hold 250 kg (550 lb) of potatoes. How much smaller is the edge of another cubic box that can hold 200 kg (440 lb)? The simple calculation is as follows. The 200-kg box has a volume that is smaller by 20 %. The volume varies as the cube of the edge—that is, the edge varies as the power 1/3 of the volume. Since 1/3 of 20 % is about 7 %, the edge of the 200-kg box is smaller by 7 %. (A more accurate calculation gives 7.2 %.)

In a final example, the power from a windmill varies approximately as the cube of the wind speed. Suppose that the wind speed increases from 12 m/s (26 mph) to 13 m/s (29 mph). How much does the power increase? The answer is, of course, a factor of $(13/12)^3$, but if we only want an approximate but quick answer, we can do the following simple calculation. Thirteen is roughly 10 % larger than 12. Since the power increases as the cube of the wind speed, going from 12 m/s to 13 m/s will increase the power by about $3 \times 10\,\% = 30\,\%$. Such crude calculations can give a useful estimate, for instance, as a simple check of data. There is no need to reach for paper and pencil or a calculator. Therefore, the method is in the arsenal of "tricks of the trade" among engineers and scientists.

By the way, $(13/12)^3 \approx 1.27$, so our approximate approach was quite reasonable.

The rule of thumb we have applied can be expressed in mathematical terms as follows:

If $Q \sim q^p$, and q changes by x %, then Q changes by $x \cdot p$ %.

The symbol $\sim$ here means "is proportional to." The number p need not be an integer, or positive. The values of p in our examples are -2 for BMI, 1/3 for the potato box, and 3 for the windmill. The rule is only approximate (mathematicians would say that it gives the first term in a series expansion), and it works best for small x, say less than 10 %.

The Aeolian Harp

Explain a phenomenon to the rest of the class in scientific terms! This is a popular pedagogical approach, even in higher education. Classmates are supposed to listen to the talk and ask questions. To the disappointment of many teachers, however, it often turns out that the student who speaks has merely looked up some facts, perhaps on the Internet, and doesn't understand much. The listeners soon get lost and are unable to ask questions. But it does not have to be like that, as the following example shows. The story is fictional but has a serious message.

A student was asked to explain the Aeolian harp, an old musical instrument whose strings are set in motion by the wind (fig. 7.3). In a rather technical presentation, the student said that as the wind blows past an obstacle, there will be a regular pattern of eddies in the air, called the von Kármán vortex street. This effect causes the wires to vibrate like violin strings, and a sound is produced. There is even a very simple formula for it, the student explained, and wrote on the board:

$$f = \frac{1}{5} \cdot \frac{d}{v}.$$

Fig. 7.3. Aeolian harp in the castle of Baden-Baden, Germany

He went on to say that f is the frequency you hear, v is the wind speed, and d is the diameter of the wire. The factor $1/5$ is called the Strouhals number. At this point, one of the students, Mary, raised her hand and asked, "Isn't the formula wrong?" The student looked bewildered at what he had written, but the teacher came to his rescue and said, "I think that changing it to this will give the correct form":

$$f = \frac{1}{5} \cdot \frac{v}{d}.$$

After the class, Mary's friend asked: "How did you know that the formula was wrong? I had never heard about von Kármán and Strouhals before." "Nor had I," said Mary. "But I know some tricks—how one can ask questions about formulas even if one has never seen them before."

Mary had applied the two most useful tricks of the trade that professional scientists and engineers use when they deal with formulas — checking the physical dimension and checking special cases and extreme limits. Many mistakes can be avoided if dimensions are checked in SI units. In this example, the wind speed is in meters per second, and the string diameter is in meters. In the first formula above, the right-hand side has the unit

$$\frac{\text{m}}{\text{m/s}} = \text{s}.$$

This is a time, but we wanted a frequency. Frequencies are commonly expressed in hertz (Hz), which is another name and symbol for the reciprocal of time, 1/s. But even if we were uncertain about this relation between hertz and second, it is obvious that a frequency cannot be expressed as a time, so the formula must be wrong anyway. The expression suggested by the teacher, on the other hand, has the correct dimensional form.

In the application of the second trick — considering special cases and extreme limits — Mary thought about what would happen if the wind speed, v, increased. In the first formula, a large v means a low frequency, f. That would be contrary to the experience that a humming sound increases in pitch if it is blowing faster, for instance, past utility lines. That is another argument showing that the student's formula was incorrect.

One Trick and Two Areas

Scientists and engineers are trained to use mathematics. Nevertheless, they often make mistakes. It is understandable that there can be errors in long and elaborate calculations, but remarkably often, the mistakes occur in very simple expressions. Perhaps we are less careful when the mathematical steps seem trivial. The inexperienced student tries to check results by going through the calculations over and over again. That is not a good strategy. If the brain makes a slip, there is a good chance that the same slip will be repeated and the

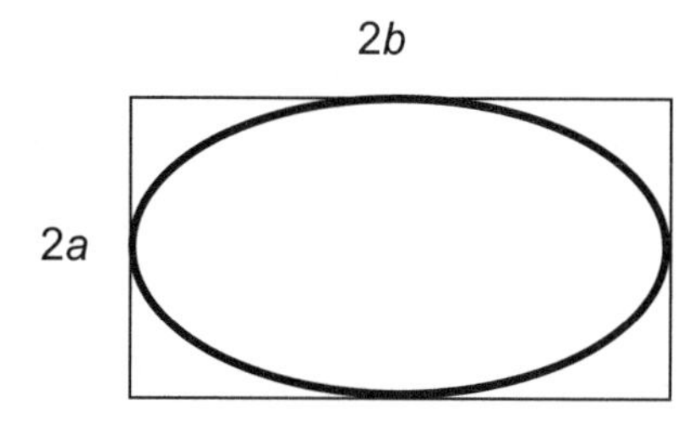

Fig. 7.4. An ellipse becomes a circle when $a = b$

error will go undetected. The professional scientist or engineer uses other tricks. One of the most important is to consider a special case. We can illustrate that approach with two simple examples from geometry—the area of an ellipse and the area of a torus.

An ellipse is characterized by the lengths a and b of its two semi-axes (fig. 7.4). The area of the ellipse is

$$A = kab.$$

Here k is a constant. For the sake of the argument, we assume that we are not sure of the value of k. It could be that we suspect a misprint in our source of the formula. How can we find the value of k, given that the expression for the area is otherwise correct? There is an obvious special case to consider, namely $a = b$, which gives a circle. Since we know for sure that the area of a circle with radius a is πa^2, the constant must have the value $k = \pi$.

A torus is a ring-shaped object like a doughnut or the inner tube in a bicycle tire. Figure 7.5 shows the geometry viewed toward the "hole" in the torus. It can be regarded as a tube of radius r that is bent in a circle of radius R. The area of a torus has a surprisingly simple mathematical form. It can be written

$$A = kRr.$$

As in the first example, we assume that the formula is correct, but we are not sure what value the constant k has. It can be determined if we consider a special case, where the area is known, or is trivial to calculate.

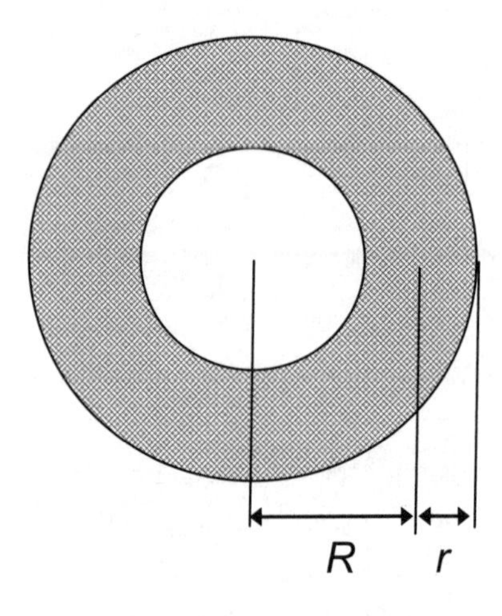

Fig. 7.5. Two-dimensional view of a torus

Let the radius of the tube, r, shrink so that it is very small but still finite. Then the torus has very nearly the same area as that of a very thin cylinder of radius r and length $2\pi R$. Since the circumference of the thin cylinder is $2\pi r$, its area is $(2\pi R)(2\pi r) = 4\pi^2 Rr$. The constant is thus $k = 4\pi^2$.

Instead of making the radius, r, very small, we could have chosen to make the radius of the torus, R, very large. That would lead to the same calculation. When we say that "r is small," we must compare r with something—in this case with R. A small r means that $r \ll R$. Similarly, a large torus radius, R, means that $R \gg r$, which is the same inequality. One cannot say exactly *how* small r has to be in an inequality like $r \ll R$. It should be "sufficiently small," and the meaning of that depends on the accuracy required in our description.

EPILOGUE

Seven Principles in Scientific Literacy

Numbers and measures are key elements in scientific literacy. This chapter looks back at many of the examples in the book and also adds some new aspects. In particular, it deals with how the language of science and technology differs from that of our daily lives—sometimes being more precise and sometimes stressing uncertainties and approximations.

It Is Not a Global World

The world of science and technology is international. English has become the *lingua franca* in scientific reports. The symbols for SI units (e.g., J for joule) and the symbols for the chemical elements (e.g., Fe for iron) are universally adopted, even in text with, for instance, Chinese characters. In daily life, however, there are national exceptions, and these can have more severe consequences than being just annoying. The ways of writing dates are so varied that one is often in doubt what day is meant. Three countries—the United States, Liberia, and Burma (Myanmar)—still use nonmetric units for length and mass, although SI units are widely used in the last two countries. We have seen how the old units caused the loss of a spacecraft to Mars and almost led to an aircraft disaster.

Further, words can have different meanings. A hurricane in the Atlantic Ocean is called a super-cyclonic storm in the North Indian Ocean, a severe tropical cyclone in Australia and the southwestern Pacific, and a super-typhoon in the northwestern Pacific. Many misunderstandings arise because of the two possible interpretations of the word *billion*. Is it one thousand million or one million million?

Those who have bought clothes in another country may have found that there is no universal system for sizes—not even in a common market like the European Union. The octane rating of gasoline depends on which standardized test procedure is used. Technical equipment sold in different parts of the world can be incompatible. For instance, North America uses 120 V, 60 Hz for domestic electricity, while most other parts of the world use 230 V, 50 Hz.

Organizations of various kinds have worked hard to eliminate national differences because they are obstacles to technical development and global commerce. It may take a long time before such goals are achieved. Until then the best we can do is to be aware of the differences.

Man Is Not Everything's Measure

"I am late because there was a long line at the post office, and the package is so heavy that I had to walk slowly," someone says. What is meant by "late," "long," "heavy," and "slowly"? In this case we compare the speaker's descriptions with our own experiences and anticipations, often in a rather vague way. It is not the length of the line in meters or yards that comes to mind, but perhaps how long it will take to reach the front of the line. In science and technology, words like *heavy* and *small* are defined by comparison with something else that has the same physical dimension (unit). We may think of our own height when we say that water is shallow, but for waves on the sea, the depth should be compared with the wavelength. What is shallow for a tsunami can be deep for a swell.

In science and math the concept of a "typical" or "characteristic" value may have a meaning that is different from what is used in colloquial conversation. For instance, in a scientific context the typical height of a human might be given as 1 m, and the height of a mountain as 1000 m. Such numbers can be accurate enough if we want to describe how the ecosphere has evolved on Earth, even though we know that the present-day human is taller than 1 m and that mountains vary a lot in height. On the other hand, if we design a rowboat, it would be better to say that the typical height of a person is, say, 170 cm. The characteristic magnitude of something depends on the purpose for which the information will be used.

Very large or very small numbers can be difficult to comprehend. If people are asked how much a bridge should be allowed to sag when a truck passes, they may respond that it should not sag at all. But before a bridge is opened for traffic, engineers measure the small amount of sagging to ascertain that it is according to the design. In our discussion of rocket launches, we noted that the risk of a fatality on the ground caused by the returning rocket had to be lower than 10^{-6}. We can never require that the risk be exactly zero. To give a feeling for what 10^{-6} means, we noted that it is the risk of being involved in a fatal car crash during about an hour's drive—something that we all can relate to.

Data Are Uncertain

Nothing that is measured with an instrument can be known *exactly*, in the strict meaning of the word. Even the counting of separate objects may have an uncertainty, as we discussed for the population of Austria on census day. Nevertheless, people would not hesitate to say that the train leaves at exactly 8:37 or that they were driving at exactly the speed limit of 50 km/h, because it expresses a kind of accuracy that is relevant in these cases. Everyone knows what they mean.

Sports results such as lengths and times are recorded according to strictly formulated international rules, but there are many inherent uncertainties. Variations in external conditions like wind speed and air pressure, and the necessary rounding of numbers for measured lengths and times, can play a role. It is unavoidable that an infinitesimal difference can tip the balance to a different result.

In technical specifications it is not enough to require, for instance, that something have a certain length. One must also specify the tolerance—how much the length may deviate from the nominal value. According to the international rules, the track in a sports stadium must have a length between 400.000 m and 400.040 m, and the length of the 50-m lane in swimming must be between 50.00 m and 50.03 m. When the domestic electrical output in North America is rated as 120 V, it is allowed to vary between 114 V and 126 V. Similarly, a factory can have specifications for how much the mass of sugar is allowed to vary in its 1-kg packages for consumers, since packages will not be exactly 1 kg.

The public often construes limit values for various hazards as sharp boundaries between what is harmless and what is dangerous. Such an attitude may be helpful in practical situations, but it obscures the arbitrariness by which limit values are usually set. One example is the body mass index, BMI. Not only have the thresholds for overweight and obesity varied, but the whole concept is meant to be a basis for general recommendations and not a precise measure of what is acceptable or dangerous for a certain individual.

A Model Is a Model

An important aspect of higher education in science and technology is training students to simplify a complex issue and focus on the most essential aspects. Those who are not used to this mode of thinking may find some simplifications unrealistic, or even ridiculous. How can anyone think of modeling a mouse as a sphere? Should one trust an analysis of a golf drive where both the golfer and the club shaft are ignored, and only the club head and the ball are considered? It all depends on the purpose of the model. When one seeks the optimum radiation dose in the hospital treatment of a tumor, every relevant aspect must be included as accurately as possible. In other cases the purpose of a model may be to understand the mechanism behind a certain phenomenon, as in our examples of how a party becomes noisy or how a tree could buckle under its own weight. Then a crude description is sufficient. As an example between these extremes, we can think of the wind chill index. It is neither necessary, nor possible, to model the risk for frostbite in great detail, but the prediction should be accurate enough to serve as a guide to the public.

Much modeling is done with computers, where input data result in a certain output. The intermediate steps may be unknown even to the modeler, who just uses the computer code that has been provided. That is a risky situation. The model may be very detailed, but with predictions that are sensitive to small variations in the parameters. Unless it is a routine problem, an experienced problem solver would check to see if the results are reasonably consistent with a much simpler model that allows calculations to be carried out with paper and pencil and a simple hand calculator. Finally, regardless of the complexity of a model, one must always remember that it is just a model — not the complete reality. But it can be a very good model!

There Are Limits to Growth

There has been unprecedented development in technology during the last century and a half. The energy use per capita in the industrialized countries has grown by more than an order of magnitude. The distance we travel during a lifetime has increased dramatically. Further, the population of the world is now six times larger than in the mid-1850s. These changes can often be described as exponential growth — a concept that appears difficult for many to fully comprehend. For instance, a steady annual increase by 2 % may seem to be a comfortably slow pace, but with the help of the rule of 72 we easily find that it means a doubling in 36 years, or not much more than one human generation.

Since no growth can continue forever, it is of interest to formulate mathematical models that can be used to extrapolate trends and predict the future. One such example is the S-shaped logistic curve, which starts out as exponential growth from low values and then bends over and saturates. The change in this curve occurs where the second derivate becomes zero. Many scientists believe that in the extraction of fossil fuel we have now passed such a point, where the sign of the second derivative went from positive to negative. The celebrated Moore's law about electronic chips is also exponential and must ultimately break down. However, that does not mean an end to technological progress. We noted how the need for light sources has been met by a sequence of innovations, from candle to gas light, the incandescent light bulb, the fluorescent lamp, and the light-emitting diode (LED). Similarly, there can be new and very different technical solutions satisfying the needs of information management.

When something grows or shrinks in size, the square-cube law can give much insight. If the linear size is doubled, the area increases by a factor of 4 and the volume by a factor of 8. This trivial scaling governs many natural phenomena. A mouse could not be much smaller without freezing to death. Roasting a turkey that is twice as heavy takes less than double the time. The scaling of the giants and the Lilliputians in *Gulliver's Travels* is a physical absurdity (but then one should remember that the book is meant to be a satire on English scientific society). Closely related to the square-cube law is the concept of a critical size, or critical mass. It originated from the construction of nuclear bombs but is now often used to refer to groups of people who interact and generate new ideas or in reference to the merger of companies.

Knowledge Is Provisional

Scientific work means new insight where knowledge was previously lacking, incomplete, or more or less incorrect. Our present understanding and quantitative description of complex phenomena is not something valid forever but may change—sometimes in a drastic way but more often through small steps. For instance, improved modeling of earthquakes has resulted in new magnitude scales. A better understanding of the long-term detrimental effects of noise has led to revised recommendations and regulations. This does not mean that the old results about earthquakes or noise were useless, but that we now have a better description or basis for decisions, and that further improvement is expected. The linear no-threshold (LNT) model for the risks associated with very low radiation doses illustrates a case where there are still huge gaps in our knowledge.

Theoretical reasoning can be powerful, but experiments are always the final judge. Galileo Galilei showed empirically that the old Greek philosophers were incorrect about falling bodies. In our day, we may think of the bathtub vortex, bicycle stability, and the thickening of old window panes in churches as examples of urban myths where reality does not agree with a popular theory. Even fundamental laws of nature may be revised, but in this case the modification usually refers to new, and perhaps very special, circumstances. The civil engineer and the nutrition chemist can still regard the separate conservation of mass and energy as absolutely true, in spite of what relativity theory has taught us about the equivalence of mass and energy. It is fascinating but sad that so many people dream of inventing a perpetuum mobile that would violate either the first or the second law of thermodynamics and solve mankind's need of energy.

Humans Are Human

Numbers and formulas have a prominent role in modern society. When used to describe complex phenomena, they can give an unwarranted aura of exactness that hides the role of human judgment. The limit values for various physical parameters in the work environment, as well as the scoring tables in decathlon and heptathlon, are the results of deliberations where different aspects are weighed against each other. The numbers and formulas are later codified through the decision of an organization or an assembly, but in practice they may depend on the views of a few individuals. Of course, those views can be very well founded and the result be widely accepted as the best possible.

Human activities mean that there is chance of human mistakes. They are unavoidable: the challenge is to find effective methods to discover and eliminate them. Some errors are elementary, similar to making an error in a physics exam problem. We have discussed several powerful approaches that are used by experienced problem solvers—checking the physical dimension (the units), considering extreme or simplified cases, and making estimates to see if the result is reasonable. Other ways of coping with possible errors are the application of large safety factors, such as those for elevator cables, and the introduction of redundancy, such as the use of dual independent car brakes.

There are also other, and more subtle, kinds of errors that are caused by a lack of safety routines or by simple misunderstanding in the communication between people. Modern safety studies have emphasized the responsibility at the management level for such errors, rather than blaming the individual who was identified as the immediate cause of a failure or accident. The loss of the Mars Orbiter and the emergency landing of an airliner are examples discussed in this book. Finally, we have mistakes arising because people with a strong conviction about something tend to see their expectations proven by what they observe. Galileo Galilei's argument that the tides are proof of the Earth's encircling the Sun is one example. Another is the alleged Russian submarine activity in the Baltic Sea in 1993. The sound picked up by sonar turned out to be noise made by swimming minks.

It is often said that each generation has to learn from its mistakes. It is also often said that we must learn from history, so that the same mistakes are not repeated. Both views are reflected in the examples in this book.

Notes

Much of the information referred to in this book has been gathered from official Web sites of national and international agencies, organizations, and so on. Those Web sites are easy to find using the names or acronyms mentioned in the text. References to books and journal articles are given below, as well as some remarks and mathematical expressions.

Chapter 1. Numbers

1. The text in the Bible, Genesis 11:1–9 (New International Version), reads: "Now the whole world had one common speech. As men moved eastward, they found a plain in Shinar and settled there. They said to each other, 'Come, let's make bricks and bake them thoroughly.' They used bricks instead of stone, and tar for mortar. Then they said, 'Come, let us build ourselves a city, with a tower that reaches to the heavens, so that we may make a name for ourselves and not be scattered over the face of the whole earth.' But the Lord came down to see the city and the tower that the men were building. The Lord said, 'If as one people speaking the same language they have begun to do this, then nothing they plan to do will be impossible for them. Come, let us go down and confuse their language so they will not understand each other.' So the Lord scattered them from there over all the earth, and they stopped building the city. That is why it was called Babel—because there the Lord confused their language of the whole world. From there the Lord scattered them over the face of the whole earth."

2. The International Organization for Standardization was established in 1947 in connection with the founding of the United Nations. It is the international organization for standardization in all technical areas except electrical engineering, which is handled by the International Electrotechnical Commission (IEC), founded in 1906. The ISO has played a very important role, for instance, in establishing definitions and symbols for scientific and technical quantities. However, its recommendations do not have the strong

legal force exemplified by the EU decision about dates on food packages (cf. sec. 1.1, What Is the Point?).

3. Two prefixes cannot be combined. This is important in the case of mass, where the unit kilogram (kg) already contains the prefix *kilo*. Therefore, prefixes must be combined with the unit gram, as with milligram (mg) and teragram (Tg). Of course, one can still write 1 mg as 10^{-3} kg and 1 Tg as 10^{12} kg.

4. Council Directive 80/181/EEC, amended through Directive 2009/3/EC.

5. In the other three Renard series, the corresponding factors are

$$\sqrt[10]{10} \approx 1.2591; \sqrt[20]{10} \approx 1.122; \sqrt[40]{10} \approx 1.059.$$

6. See note 16, below, for a logarithmic scale of grain sizes.

7. The need for greater accuracy has also led to scales E24, E96, and E192, where each step increases by the factors $\sqrt[24]{10} \approx 1.101$, $\sqrt[96]{10} \approx 1.024$, and $\sqrt[192]{10} \approx 1.012$, respectively.

8. The area of A0 is 1 m^2. It is then halved in the sequence A1, A2, A3, A4, A5, and so on. The figure shows the step from A3 to A4.

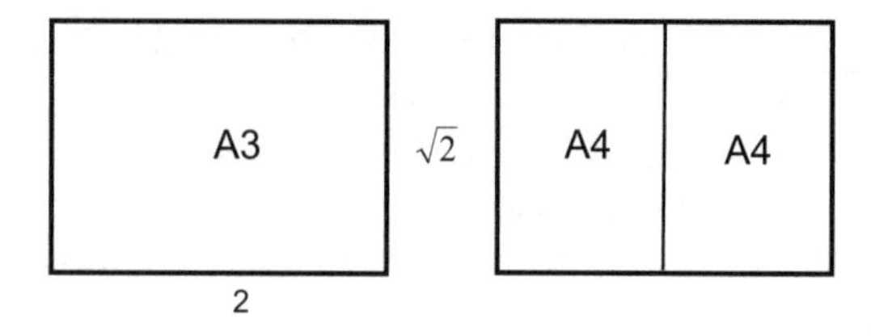

9. In tables of a quantity that spans many orders of magnitude, the probability p_n that the first digit has the value n is $p_n = \log_{10}[(n+1)/n]$. The sum of all probabilities must be equal to 1: $p_1 + p_2 + \ldots + p_9 = (\log 2 - \log 1) + (\log 3 - \log 2) + \ldots + (\log 10 - \log 9) = \log 10 - \log 1 = 1 - 0 = 1$.

10. The decreasing probability of the numbers 1 through 9 as the first digit in tables has an analogy in the *Zipf rule*, although for different reasons. The American linguist George K. Zipf studied the inverse relation between the number of letters in a word and the frequency of that word.

11. There are many heuristic ways to motivate the expression for p_n. Here is one example. Let E_i be the numerical value for the energy in a certain unit system, and write it as $E_i = e_i \cdot 10^r$, where $1 \leq e_i < 10$. The relation above for p_n can be generalized to the statement that the probability for e_i to lie between $e_{i,1}$ and $e_{i,2}$ is $p(e_{i,1} \leq e_i \leq e_{i,2}) = \log_{10}[e_{i,2}/e_{i,1}]$. In another unit

system E_i would be kE_i, where k is the conversion factor between the systems. Thus, e_i is replaced by ke_i. We now see that $p(ke_{i,1} \leq ke_i \leq ke_{i,2}) = \log_{10}[ke_{i,2}/(ke_{i,1})] = \log_{10}[e_{i,2}/e_{i,1}]$. The probability should not depend on the unit system chosen. We note in Table 1.6 that the distribution of first digits has the same character after a change in the energy unit. In fact, the logarithmic law in the frequency of the first digit is the only distribution that is invariant with a change of the unit.

12. A distribution $f(x)$ contains the same information as a cumulative distribution $g(x)$ since $f(x) = \mathrm{d}g(x)/\mathrm{d}x$.

13. Matthias R. Mehl, Simine Vazire, Nairán Ramírez-Esparza, Richard B. Slatcher, and James W. Pennebaker, "Are Women Really More Talkative Than Men?" *Science* 317 (2007): 82.

14. It is ironic that sometimes people were flown out by helicopter from the Esrange rocket test range—a safety measure that probably involved greater risks than if they had remained unprotected during the launch. The space agency also built shelters that could be used by the local Sami people, who occasionally were herding reindeer in the area. But such shelters also tended to attract outdoor people, who were not easy to reach with information about launches.

15. The probability of completely avoiding something at the risk level of p during n consecutive periods of this risk is $(1 - p)^n = \exp[n\ln(1 - p)] \approx \exp(-np) = 1/\mathrm{e} \approx 0.37$ when $n = 10^6$ and $p = 10^{-6}$. A long life is a total of about 10^6 hours. If each hour of life can be associated with an activity that has a fatality risk of 10^{-6}, the chance is approximately 37 % that one will *not* die because of these activities. One hour in a car means a risk of a fatal crash that is of the order of 10^{-6}. People normally accept this risk, driving without much thought about the obvious risks. We can conclude that most of us would readily engage in activities at this risk level, in spite of the fact that accidents and fatalities are not negligible.

16. One mole of an element (i.e., as many grams as the number which expresses the relative atomic mass, a number earlier called the atomic weight) contains about 6×10^{23} atoms (Avogadro's constant). Sand does not consist of a single chemical element, but for our purposes we can represent sand with an element having a relative atomic mass of, say, 30, and a mass density 3000 kg/m^3. A grain in the form of a cube with sides 1 mm in length has the volume 1×10^{-9} m^3 and therefore a mass of 3×10^{-3} g, corresponding to about 10^{-4} mole of atoms. This gives more than 10^{19} atoms per grain of sand. That number of grains, each with a volume $1 \times$

10^{-9} m^3, would form a sandpile with a volume larger than 10^{10} m^3, or larger than the volume of a cube with sides of 1 km (0.6 mile).

Chapter 2. Measures

1. Harry O. Wood and Frank Neumann, "Modified Mercalli Intensity Scale of 1931," *Bulletin of the Seismological Society of America* 21 (1931): 271.

2. Scott Huler, *Defining the Wind* (New York: Crown, 2004).

3. The life of Charles F. Richter is described in Susan Elizabeth Hough, *Richter's Scale* (Princeton: Princeton University Press, 2007).

4. The moment magnitude M_w is defined as

$$M_w = \frac{2}{3} \log_{10}(10^7 M_0) - 10.7,$$

where M_0 is the numerical value of the so-called seismic moment expressed in newton meters (Nm). M_0 is also a measure of the amount of released energy.

5. The official IAEA Web site uses slightly different words for INES levels 4 and 5, and also other wording to describe the three areas, but the meaning is the same as described here.

6. The four indicators for human poverty index HPI-2—P_1, P_2, P_3, and P_4—range from 0 to 100, since they are expressed in percent. A large P_i means a large deprivation for that dimension (for instance, a large fraction of the population being illiterate). HPI-2 is obtained from the four indicators as

$$\text{HPI-2} = \left[\frac{1}{4}\,(P_1^\alpha + P_2^\alpha + P_3^\alpha + P_4^\alpha)\right]^{1/\alpha}.$$

It is interesting to investigate the role of the parameter α. If $\alpha = 1$, we get a plain arithmetic average of the indicators P_i—that is, all deprivation indicators are equally weighted. If α becomes very large, HDI-2 will tend to the value of the largest P_i, so HDI-2 enhances that dimension in which there is the greatest deprivation. It has been judged that $\alpha = 3$, a compromise between these extremes, gives a reasonable account of poverty.

The poverty index HPI-1 has only three dimensions, described by the indicators P_1, P_2, and P_3. They are combined in analogy to HPI-2, but still with the choice $\alpha = 3$.

7. In the gender-related development index, one first calculates male and

female indices $P_{i,\mathrm{M}}$ and $P_{i,\mathrm{F}}$ for each dimension. They are weighted by the male and female population shares S_{M} and S_{F}, in an expression that yields the equality distributed index $P_{i,\mathrm{ED}}$ as

$$P_{i,\mathrm{ED}} = [S_{\mathrm{M}} \cdot (P_{i,\mathrm{M}})^{1-\epsilon} + S_{\mathrm{F}} \cdot (P_{i,\mathrm{F}})^{1-\epsilon}]^{1/(1-\epsilon)}.$$

If $\epsilon = 0$, we get a plain arithmetic average that takes into account the relative sizes of the male and the female population. With a very large and positive ϵ, $P_{i,\mathrm{ED}}$ is dominated by the smallest P_i (since $P_i < 1$)—i.e., it strongly emphasizes the influence of the underprivileged sex.

8. Taking the derivative dP/dR in the expression for the points, P, obtained in a decathlon event, we can write

$$\frac{\Delta P}{P} = C\left(\frac{R}{R-B}\right)\frac{\Delta R}{R}.$$

Here ΔP is the change in the number of points when the result R is changed by a small amount, ΔR. It can be given the following interpretation. If the numerical value of the achievement R in a certain event is improved by 1 % (i.e., $\Delta R/R = 0.01$), the relative change $\Delta P/P$ in the assigned number of points is $CR/(R - B)$ %. From the numbers in Table 2.4, we see that in the throwing events—shot put, discus, and javelin—C and $R/(R - B)$ are both close to 1. This scale awards points that are almost linear in the increment of the numerical value of the achievement. The three track events have C and typical $R/(R - B)$ close to 2. These differences reflect the fact that the spread in R is larger in the throwing events than in the running events. For instance, the spread $\Delta R/R$ between the first and the sixth place at the Beijing Olympics was 1.031 for the men's 400 m, 1.046 for the men's pole vault, and 1.085 for the men's javelin throw.

9. Our example was for the special case of nine teams. Similar scales can be constructed for other cases, but the idea works best when the number of contestants is not too small.

Chapter 3. Accuracy and Significance

1. The accuracy of a measurement refers to how close the measured value is to the true or accepted value. The uncertainty of a measurement describes the region about an observed value within which the true or accepted value probably lies.

2. The story about the New York subway uses some data from Thomas L. Bohan, *Crashes and Collapses* (New York: Checkmark Books, 2009).

3. There has been much discussion about whether Armstrong used the word "a" before "man." A computer analysis of the recorded sound by Peter Shann Ford, Sydney, suggested that there is a very short "a," although the communication technology of 1969 made it practically inaudible. There is no consensus, however, on this interpretation.

4. The experiment is described in J. C. Hafele and R. E. Keating, "Around-the-World Atomic Clocks: Predicted Relativistic Time Gains," *Science* 177 (1972): 166. The analysis must account for the fact that the stationary clock on the Earth is not in an inertial frame, since the Earth rotates. Further, there is an effect due to the flight altitude, which changes the gravitational field.

5. R. Matthews, "Storks Deliver Babies ($p = 0.008$)," *Teaching Statistics* 22 (Summer 2000): 36; S. Wirth, "King Kong, Storks, and Birth Rates," *Teaching Statistics* 25 (Spring 2003): 29.

6. If n is the number of votes in a sample, and p is the fraction of votes that an alternative gets (corresponding to $100p$ % of the votes), the uncertainty in p for that alternative is

$$1.96\sqrt{\frac{p(1-p)}{n}}\ .$$

Table 3.3 is based on this convention to express margins of error.

7. G. G. Luther and W. R. Towler, "Redetermination of the Newtonian Gravitational Constant G," *Physical Review Letters* 48 (1982): 121; M. P. Fitzgerald and T. R. Armstrong, "Newton's Gravitational Constant with Uncertainty Less than 100 ppm," *IEEE Transactions on Instrumentation and Measurement* 44 (1995): 494; P. J. Mohr and B. N. Taylor, "CODATA Recommended Value of the Fundamental Physical Constants: 2002," *Reviews of Modern Physics* 77 (2005): 43; P. J. Mohr, B. N. Taylor, and D. B. Newell, "CODATA Recommended Value of the Fundamental Physical Constants: 2006," *Reviews of Modern Physics* 80 (2008): 687. There are other recent measurements claiming high accuracy.

8. A discussion of the effect of wind and altitude is given by H. Frohlich, "Effect of Wind and Altitude on Record Performance in Foot Races, Pole Vault, and Long Jump," *American Journal of Physics* 53 (1985): 726. The well-known formula $s = v^2/g$ for the maximum distance s reached by a projectile that is launched with the speed v is too simple to be applied to long jump. Frohlich argues that the effect of g on the jump length varies as $1/g^{1/2}$.

Chapter 4. Extrapolations

1. In our discussion of exponential growth, we have ignored the difference between discrete growth, where the increase occurs at certain equidistant times (e.g., the number of participants in an annual event), and a continuous growth. Let $Q(t)$ be a "production" of some kind, which grows continuously with time, t, according to $Q(t) = Q_0\exp(kt)$, and τ be the doubling time, i.e., $\exp(k\tau) = 2$. Then

$$\int_{-\infty}^{t+\tau} Q(t)\,dt = \int_{-\infty}^{t} Q(t)\,dt + \int_{t}^{t+\tau} Q(t)\,dt = (Q_0/k)\exp(kt)$$
$$+ (Q_0/k)\exp(kt) = 2(Q_0/k)\exp(kt).$$

This is the rule about the accumulated amount. In the second integral, we used $\exp(k\tau) = 2$. The fact that we integrated from a time $-\infty$, rather than from a finite time, is not important because the area under the initial "tail" is small when $t \gg \tau$.

2. The curve in Fig. 4.1 has the mathematical form

$$\frac{1}{1 + (\pi^2/400)t^2},$$

where t is the time in years. An integration over t from 0 till infinity gives the value 10—i.e., the same as the integration over the thick-line distribution in Fig. 4.1.

3. Brian J. Whipp and Susan A. Ward, "Will Women Soon Outrun Men?" *Nature* 355 (1992): 25.

4. Raymond Kurzweil, "The Law of Accelerating Returns," at *www.kurzweilai.net*, 7 March 2001.

5. Galileo Galilei, *Dialogues Concerning Two New Sciences* (1638), trans. Henry Crew and Alfonso de Salvio (New York: Macmillan, 1914).

6. The logistic function (logistic curve) can be written

$$N = \frac{N_0}{1 + e^{-t}}.$$

We may think of N as a population (expressed in a suitable unit) and t as time. The limit value N_0 after long times may be called the carrying capacity, i.e., an assumed maximum population, given the available resources. A mathematically equivalent form is $N = N_0\,e^t/(e^t + 1)$.

7. Two other common sigmoid-like expressions are the Gompertz growth function, $N = N_0\exp[-\exp(b - ct)]$, with an inflection point at $t =$

b/c; and the logistic or Verhulst growth function, $N = N_0/[1 + b\exp(-ct)]$, with an inflection point at $t = (\ln b)/c$.

8. Charles J. Krebs, Rudy Boonstra, Stan Boutin, and A. R. E. Sinclair, "What Drives the 10-Year Cycle of Snowshoe Hares?" *BioScience* 51 (2001): 25.

9. The most common chemical clock uses the Briggs-Rauscher reaction. Three colorless solutions are mixed. Then the mixture starts oscillating between being clear and deep blue, with a period of several seconds. After several minutes the oscillations have died out, and the solution ends up as a dark mixture. The phenomenon can be accurately described by a mathematical model with coupled differential equations.

Chapter 5. Models

1. This model is given in George Polya, "Probabilities in Proofreading," *American Mathematical Monthly* 83 (1976): 42.

2. William Thomson, "On the Secular Cooling of the Earth," *Philosophical Magazine*, ser. 4, 25 (1863): 1; Joe D. Burchfield, *Lord Kelvin and the Age of the Earth* (Chicago: University of Chicago Press, 1990).

3. W. R. MacLean, "On the Acoustics of Cocktail Parties," *Journal of the Acoustical Society of America* 31 (1959): 79.

4. People modify their speech in the presence of disturbances (the Lombard effect), so it is not quite correct to assume a constant power, P. However, that does not affect the argument here.

5. Some research on the "cocktail party effect" has dealt with contributions to the signal other than just sound. A listener looking at the speaker considerably enhances the signal by interpreting the speaker's body language. In some cases lip reading is a helpful clue. Next time you are at a loud party, try this experiment: close your eyes and find out if you can still follow your neighbor's conversation.

6. To describe the electrostatic forces expressed by Coulomb's law as mediated by photons, we can write the force between the charges that varies with the distance r as

$$\frac{1}{r^2}\,(1 + \mu r)e^{-\mu r}.$$

Here, $\mu = 2\pi m_0 c/h$ in an inverse length, where c is the speed of light in vacuum and h is Planck's constant. If the photon rest mass $m_0 = 0$, we get exactly the r^{-2} law. If $\mu r \ll 1$, Coulomb's law with the exponent -2 is valid to a high accuracy. A series expansion gives the first two terms in the expression above as

$$\frac{1}{r^2}\left(1 - \frac{1}{2}(\mu r)^2\right).$$

Astrophysical measurements show that $\mu < 10^{-24}\ \mathrm{m}^{-1}$. The condition $\mu r \ll 1$ then implies that the r^{-2} law is a very good approximation at least up to distances $r \sim 10^{24}$ m. Further, the limit $\mu < 10^{-24}\ \mathrm{m}^{-1}$ implies that $m_0 = \mu h/(2\pi c) < 5 \times 10^{-63}$ kg. Compare this with the mass of an electron, 9.1×10^{-31} kg.

7. The Faraday's cage effect is demonstrated in some science museums. The cage itself is usually formed as a sphere made of a metallic net. A museum employee opens a door in the sphere, steps inside, and closes the door. The cage, hanging on an electrically isolated cable, is hoisted into the air. The cage is hit by an electric discharge, like a lightning bolt, from a high-voltage generator. Then the cage is lowered to the floor again, the employee opens the door, and steps out unhurt. It is the same phenomenon that makes a car a safe place during a thunderstorm.

Chapter 6. The Real World

1. The popular concept of centrifugal force is not formally correct, and we should instead introduce a centripetal force, in the opposite direction.

2. David E. H. Jones, "The Stability of the Bicycle, *Physics Today*, April 1970, pp. 34–40. Reprinted in *Physics Today*, September 2006, pp. 51–56.

3. Peter Richardson, "It Is as Easy as Falling Off a Bike . . . " *New Scientist*, April 30, 1987, p. 36, describing an attempt by Tony Doyle, of Sheffield University, to build a destabilized bicycle where the castor effect has been eliminated.

4. C. P. Saylor, "Case of the Flowing Roof," *Chemistry* 44 (December 1971): 19.

5. A. H. Shapiro, "Bath-Tub Vortex," *Nature* 196 (1962): 1081 (Massachusetts); A. M. Binnie, "Some Experiments on the Bath-Tub Vortex," *Journal of Mechanical Engineering Science* 6 (1964): 256 (England); L. M. Trefethen, R. W. Bilger, P. T. Fink, R. E. Luxton, and R. I. Tanner, "The Bath-Tub Vortex in the Southern Hemisphere," *Nature* 207 (1965): 1084 (Australia).

6. Trefethen et al., "Bath-Tub Vortex in the Southern Hemisphere."

7. A discussion of Galileo Galilei's arguments is given in Hans C. Ohanian, *Einstein's Mistakes* (New York: W. W. Norton, 2008).

8. The present author was a member of this commission. The detailed and unclassified report (*Ubåtsfrågan, 1981–1994*, Statens Offentliga Utredningar, SOU 1995:135) has a summary in English.

9. In the square net, we can distinguish between bond percolation and site percolation. In the former case, we cut individual links (bonds), and in the latter case we randomly remove the nodes (sites) where four links meet. The bond percolation limit is 0.5, and the site percolation limit is approximately 0.59 (i.e., the net falls apart when 41 % of the sites are removed).

10. Let the weight of the tree crown be represented by the force, P, applied a small horizontal distance, ϵ, from that position, which is exactly above the center of the "root." The weight of the column itself is ignored. The elastic energy stored in the bent column depends on how flexible it is. This is determined by three quantities: the elastic modulus, E, of the material; the second axial moment of inertia, I; and the length, L. In equilibrium, when the elastic energy balances the change in gravitational potential energy, elasticity theory shows that the top of the column bends out a horizontal distance δ, which is calculated as

$$\delta = \epsilon \frac{1 - \cos\sqrt{PL^2/(EI)}}{\cos\sqrt{PL^2/(EI)}} .$$

Because $\cos(\pi/2) = 0$, we see that δ becomes infinite when the load, P, approaches a critical value P_c, with

$$P_c = \frac{\pi^2 EI}{4L^2},$$

irrespective of how small ϵ is. If $\epsilon = 0$ (exactly), the horizontal displacement would always be zero, but that is merely a mathematical idealization that would not occur in nature.

11. In thermodynamics, it is not the energy but the free energy (Gibbs energy or Helmholtz energy) that will attain its lowest value in equilibrium.

12. Kurt Vonnegut, *Slapstick; or, Lonesome No More!* (1976; New York: Bantam, 1989), chap. 31.

Chapter 7. Tricks of the Trade

1. An iron-nickel alloy with very high magnetic permeability, called μ-metal (mu-metal), gives a good, although not perfect, shield against magnetic fields.

2. The floating apple, deformed to a slab in the figure below, is submerged the distance a in oil and b in water, with x being the distance from the top of the slab to the liquid. If the densities are ρ_{apple}, ρ_{oil}, and ρ_{water}, we get

$$(a + b + x)\rho_{\text{apple}} = a\rho_{\text{oil}} + b\rho_{\text{water}}.$$

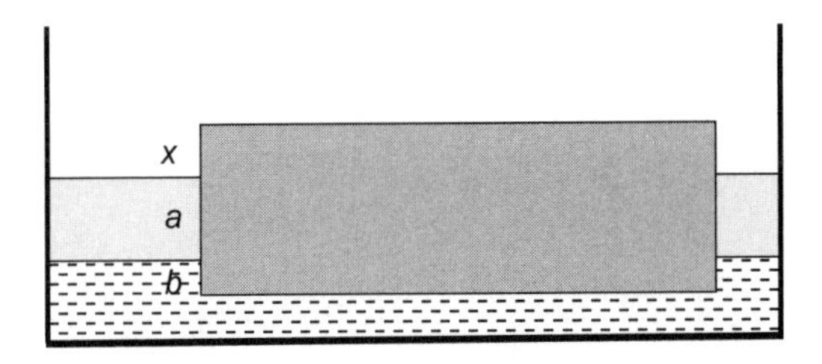

3. If the temperature at the surface varies as $T_s(t) = T_o + \Delta T \sin(\omega t)$, the differential equation for heat flow yields the temperature $T(x,t)$ at depth x and time t:

$$T(x,t) = T_o + \Delta T \exp\{-x\sqrt{a/(2a)}\}\sin\{\omega[t - x\sqrt{1/(2a\omega)}]\}.$$

The amplitude of the temperature variation thus decreases exponentially with the depth x. It is reduced by a factor $1/e$ at $x = L\sqrt{2}$, where $L = \sqrt{a/\omega}$. The maximum temperature at depth x is delayed in time by $x\sqrt{1/(2a\omega)}$. The angular frequencies in our examples are $\omega = 2\pi/T_{\text{year}}$ and $\omega = 2\pi/T_{\text{day}}$, respectively.

4. The motion of a falling body of mass m subject to a retarding force $F = \frac{1}{2}C_D\rho A v^2$ due to air drag is discussed in detail in Göran Grimvall, *Brainteaser Physics* (Baltimore: Johns Hopkins University Press, 2007). Newton's equation of motion gives the distance s after time t as

$$s = a \ln\left[\cosh\left(t\sqrt{\frac{g}{a}}\right)\right],$$

where $a = 2m/(C_D A\rho)$ and g is the acceleration of gravity.

Index